The Legacy of Botanical Latin

Also available from Bloomsbury

Analysing the Boundaries of the Ancient Roman Garden by Victoria Austen
Ancient Magic and the Supernatural in the Modern Visual and Performing Arts edited by Filippo Carlà-Uhink and Irene Berti
Magic in Ancient Greece and Rome by Lindsay C. Watson
Mountain Dialogues from Antiquity to Modernity edited by Dawn Hollis and Jason Konig

The Legacy of Botanical Latin

On Naming and Knowing Plants and the Issue of Universal Intelligibility

Erin Petrella

BLOOMSBURY ACADEMIC
LONDON • NEW YORK • OXFORD • NEW DELHI • SYDNEY

BLOOMSBURY ACADEMIC
Bloomsbury Publishing Plc, 50 Bedford Square, London, WC1B 3DP, UK
Bloomsbury Publishing Inc, 1359 Broadway, New York, NY 10018, USA
Bloomsbury Publishing Ireland, 29 Earlsfort Terrace, Dublin 2, D02 AY28, Ireland

BLOOMSBURY, BLOOMSBURY ACADEMIC and the Diana logo are trademarks of Bloomsbury Publishing Plc

First published in Great Britain 2026

A catalogue record for this book is available from the British Library.

A catalog record for this book is available from the Library of Congress.

ISBN: HB: 978-1-3505-0966-5
ePDF: 978-1-3505-0967-2
eBook: 978-1-3505-0968-9

Typeset by RefineCatch Limited, Bungay, Suffolk

For product safety related questions contact productsafety@bloomsbury.com.

To find out more about our authors and books visit www.bloomsbury.com and sign up for our newsletters.

ad fratrem (to my brother)

Contents

Figures

Preface

This book is the result of many years of reflection. On the surface, it is a standard conversion of dissertation into monograph, but its origins far pre-date my graduate work. A lifetime ago, I was a gardener in New York City, working first in the neglected parks of Washington Heights and later for a company with a variety of uninspiring commercial clients. I had studied Latin in secondary school and forgotten it entirely, but soon became fascinated with the scientific, Latin names of the plants with which we worked. In that world, where crew assignments are based on skill level, this kind of knowledge did not simply afford one bragging rights. It translated directly into better pay and easier, less labor-intensive work. I eventually left gardening and found myself relearning Latin, at which point I realized that there was more to botanical Latin than fancy nomenclature. As I worked my way toward a PhD in Classics, I continued to ruminate on botanical Latin, even when inundated with Homer, Virgil, and Cicero. I have always been drawn to the peripheral, to fringe people and ideas, and I was fixated on the fact that, in contrast with the standard, inanimate *materia* of Classics, botanical Latin was still alive and perhaps not well, perhaps merely languishing, but surviving nonetheless. There was a vibrant, ongoing debate about botanical Latin and I was determined to see how it played out.

From a classification standpoint, this book is fundamentally interdisciplinary. It is about the history and development of a hyper-specific use of the Latin language that developed in the Early Modern era, but it is also about the philosophical and moral implications of the current debates in the field of botany. The overall structure reflects these twin projects, with the first three chapters tracing the origins of botanical Latin in antiquity through the Renaissance and Early Modern eras. In Chapter 4, the discussion turns to the nineteenth and twentieth centuries, when botanical Latin rapidly developed into its modern iteration as a scientific language far removed from its origins as a natural language in antiquity. From this, a broader discussion opens up about the role of botanical colonization, the moral-ethical implications of Latin for botanical nomenclature, and the efficacy of visual forms of knowledge and information transmission. The text concludes with a critique of the scientific ideals of universal intelligibility, discoverability, and knowability. The first of these, universal intelligibility, is the assumption that it is possible to be understood by anyone trained in the field of botany, by means of strictly regulated nomenclatural and descriptive rules, one of which has traditionally been the use of the Latin language. By discoverability, I mean the extent to which the representation of a plant, whether textual, oral, or visual, aids in its discovery outside of the context of its naming and description. Knowability refers to the degree to which, and the contexts within which, a plant can be said to be known.

Given the critical nature of this discussion, I find it fitting to also address the relation of this book to the field traditionally known as Classics, usually defined by the study of Greek and Latin languages and literature from antiquity to roughly the Renaissance. This characterization is sometimes supplemented by social and historical analysis or, in recent decades, reception studies and Neo-Latin. Scientific Latin, however, especially post-Renaissance, is rarely studied in-depth, which is unfortunate, given the influences of ancient writers such as Theophrastus, Dioscorides, and Pliny the Elder on the field of botany. In addition, the field has a long history of elitism, racism, misogyny, and classism, and frequently ignores the reception of ancient literature and culture by non-Western scholars. Many departments have begun referring to the field as Ancient World Studies, a shift that expands its limits to include other languages and cultures and more inclusive genres than philology and material and literary analysis. A new methodology known as Critical Ancient World Studies (CAWS) actively engages with the field's colonial (and present) history. In particular, CAWS is the process of forgetting, a rejection of the study of antiquity as a reconstruction of European culture and ideals. Similarly, I see the current debates in botany as evidence of a crisis the resolution of which lies not in an attempt to continue the discourse within the usual scientific framework, but in the recognition that science does not exist in a vacuum and cannot be divorced from ethical concerns. Botany's colonial past (and the use of Latin for naming and description is a part of that history) has shaped the field, resulting in a hierarchy in which many non-Western practitioners and non-professionals feel excluded. Scientific Latin is just as fascinating as epic poetry and reveals just as much to us about ourselves as elegy and philosophy. This is the world in which we live, this is the Latin (and Greek) that surrounds us. Scientism, the assumption that knowledge is defined almost exclusively by what can be and has been subjected to formal scientific investigation, is our intellectual milieu, the standard to which all discourse is held.

However, I want to make a further point, aimed at scholars of antiquity: the study of ancient Greek and Latin literature, philosophy, science, and history is directly tied to the ways in which our contemporaries appropriate those same genres. Scientific Latin may not be "literary", but it has had a profound impact on contemporary scholarly discourse, which has itself trickled down into our everyday vernacular. It has become the *status quo* in online discussions to demand citations in support of even the most banal of claims. These are the effects of scientism and centuries of conditioning on how to approach the acquisition of knowledge. Latin (and to a lesser extent Greek) has not just been implicated but fully involved in the development of this concept of knowing. In the world of gardening, promotion is often tied to learning the scientific names of plants, under the assumption that one does not know them unless one also knows these names. The task of a gardener is not, however, to rank a plant taxonomically, but to know its ideal growing conditions, soil needs and remediation, pest control, and pruning techniques, none of which is dependent on the scientific name. Given this, and in light of current debates in botany, botanical Latin can be seen as a study in support of CAWS. Forgetting Classics does not just mean expanding the limits of what is considered classical or ancient literature; it also involves critiquing the legacies of the two traditional classical languages, Greek and Latin. In botany, critique of the legacy of Latin involves the scientific ideals that have historically been embedded in it.

Acknowledgments

Many people have supported and encouraged me in the writing of this book. First and foremost, I am indebted to my family and my friends Corrie, Jerah, and Christine for tolerating my idiosyncrasies. I also owe much to Carmela Franklin, who has supported me for over a decade and taught me everything I know about manuscripts. I am also grateful for the guidance of Gareth Williams and Joseph Howley, Pamela Smith, Benjamin Fortson, and Alex George, all of whom have given me excellent information and advice that has changed the trajectory of this project. The Classics department at Columbia University taught and shaped me, especially Liz Scharffenberger, who first convinced me to do a postbac in Classics. I am also grateful for the many conversations I have had with Brett Stine, Umberto Verdura, Shanté Tucker, José Antonio Cancino Alfaro, and others. Portions of this project have also been presented at the Brooklyn Institute for Social Research, the Society for Classical Studies, and the NeoLatina conference at the Universität-Freiburg. I am grateful to everyone who organized, participated in, and attended these talks. I have incorporated many of their suggestions and greatly appreciate the opportunity to present my thoughts to such wonderful people. For making their images readily available, I am indebted to the Wellcome Library and to Penn University's OPenn project. In the Spring of 2025, I had the privilege of attending an Interdisciplinary residency at Oak Spring Garden Foundation, whose library collection supplied the majority of the botanical illustrations in Chapter 5. The library staff, headed by Tony Willis, were accommodating and knowledgeable about the history of plant illustrations. I am also indebted to the botanists in residence, who humored my ideas and gave me invaluable professional feedback: Cade Kane, Israel Borokin, Michael D'Antonio, and Rafael Urbina-Casanova. Thanks to everyone in the Bellflowers cohort and the staff who supported us, Goldie Poblador and James Adelman in particular.

Note on Translations

All translations are the author's except where noted.

Abbreviations

CB	*Critica Botanica* (Linnaeus, 1737)
CP	*De Causis Plantarum* (Theophrastus)
DOGO	*De Optime Genere Oratorum* (Cicero)
FB	*Fundamenta Botanica* (Linnaeus, 1736)
GP	*Genera Plantarum* (Linnaeus, 1743)
HN	*Historia Naturalis* (Pliny the Elder)
HP	*Historia Plantarum* (Theophrastus)
IBC	International Botanical Congress
ICBN	International Code of Botanical Nomenclature (obsolete)
ICN	International Code of Nomenclature for algae, fungi, and plants
IRBN	International Rules of Botanical Nomenclature (obsolete)
ONF	*On the Natural Faculties* (Galen)
SP	*Species Plantarum* (Linnaeus, 1753)
PB	*Philosophia Botanica* (Linnaeus, 1751)

Introduction

In 1743, Carl Linnaeus (1707–78), while discussing his work on the formalization of plant classification, defined a botanist as anyone who "knows how to assign similar names to similar plants, and distinct names, which are intelligible to everyone (*cuicunque intelligibilibus*), to distinct plants."[1] The botanist is therefore someone who has learned how to classify and name plants by organizing them according to common features. Embedded in this statement is the concept of universal intelligibility, the idea that when one is describing, classifying, and naming plants, it is possible to be understood by anyone. Until the early twentieth century, many botanists assumed that the formalization of botanical Latin and nomenclature would render genera and species from the plant kingdom knowable on an international level. However, by the mid-twentieth century, proficiency in Latin had become an anomaly rather than a given and the vast corpus of untranslated botanical Latin literature, dating back centuries, had become unintelligible to many botanists.

By the end of the twentieth century, botanical Latin was in most cases used only because the International Code of Nomenclature for algae, fungi, and plants (ICN), generally referred to by practitioners as the "Code," required it.[2] Prior to the 1930 Cambridge International Botanical Congress (IBC),[3] the use of Latin in botany, while not mandatory, was readily used. In 1935, the IBC instituted a rule stating that diagnoses or descriptions of new botanical organisms must be composed in Latin, which instantaneously became a necessary tool for communication between botanists around the world. For European botanists up through the nineteenth century, Latin was the *lingua franca*, the language that scholars were expected to use in correspondence and in their published works. For the modern botanist, this had not been the case for a long time, partially because of the waning of Latin as a common language, but also because many botanists are non-European and speak languages that have little or nothing to do with Latin.[4] At the same time, the vocabulary of botany had become more complex and less accessible to laypeople. After 1935, fewer and fewer modern botanists knew Latin well enough to compose and read formal botanical diagnoses. As a result, examples of botanical Latin from the twentieth century and later tend to be simplistic in both grammar and syntax, with highly technical vocabulary. This was the context in which William Stearn published *Botanical Latin* in 1966, a manual that explained for botanists the long-forgotten origins of their "international" code of communication and how to use it.[5] Stearn's monograph was aimed at botanists with little to no knowledge of classical Latin, which differs from botanical Latin in the latter's need for "precision and economy in words."[6] Stearn's project did not involve extensive analysis of the origins of botanical Latin. Rather, his purpose was to explain the basics of the grammar and syntax of the language to non-Latinists, along with the idiosyncrasies of scientific descriptions of plant species, so that botanists could read and understand not only modern descriptions but older ones as well. A large proportion of the text is focused on the language of botany and it served for decades as the definitive guide for reading and composing formal botanical diagnoses for publication, in accordance with the ICN.[7]

In the last few decades of the twentieth century, the ICN's Latin requirement for diagnoses was heavily debated,[8] but it was not until 2011, at the Eighteenth IBC in Melbourne, that the decision was made to amend the 1935 Latin requirement. The result of the Melbourne Congress was a new ICN, put into effect on January 1, 2012. Known also as the Melbourne Code, the ICN dictates that the names of organisms must still be devised in Latin and must still adhere to Linnaeus' binary rules of nomenclature.[9] However, diagnoses may now be published in either Latin or English. With Latin no longer the *lingua franca*, centuries of botanical Latin diagnoses and descriptions have been relegated to the annals of history. This provided the basis for Emma Short and Alex George's 2013 *Primer of Botanical Latin*,[10] a manual for schooling botanists on how to compose and translate botanical diagnoses. In 2020, Ross Bayton published *The Gardener's Botanical*,[11] a compilation of definitions of Latin plant names with full-color illustrations. Peter Jones, whose 2016 *Quid pro Quo* discusses the etymology of a variety of Latin terms in English, including in botany, cheekily proclaims *nil desperandum* (don't despair) about the 2012 Code.[12] Lorraine Harrison's 2012 *Latin for Gardeners* promises to provide to its readers an experience similar to traveling overseas with a basic grasp of another language,[13] while John Wright's delightful 2014 *The Naming of the Shrew* guides the reader through the history and idiosyncrasies of scientific nomenclature, without which "chaos would rule the science of biology."[14] Such works are not mere novelties. They are necessary to open up the world of botanical Latin to specialists and non-specialists alike, unlocking its secrets and exposing the information hidden in it. The publication of such manuals is evidence that botanical Latin as a distinct linguistic entity[15] has enduring professional value. There are, however, very few scholarly studies of its origins as a formal, scientific tool that might answer the questions: How did botanical Latin develop and in what context? And, why did botany, of all of the scientific disciplines, develop such stringent rules for the expression of its material?[16] While the answers to these questions are complex, there are several identifiable pieces of this highly detailed historical-linguistic puzzle dating back to antiquity. There are also lingering questions regarding the legacy of botanical Latin, which are complicated by serious ethical debates that have sprung up in the past decade and are fundamentally tied to the field's colonial history and to its attachment to the Latin language.

Botanical Latin

Botanical Latin is not just the set of botanical texts that happen to have been composed in Latin, nor, despite its obvious deviation from classical and medieval Latin, was it ever a dead language. It was a constantly evolving, highly technical, structured code, the systematization of which took centuries to develop. By the twentieth century, the term "botanical Latin" no longer referred to a natural language, but to a system of recommendations for the composition of formal diagnoses of new genera and species, with a solid history of common syntactical constructions and a highly specialized vocabulary.[17] Perhaps the most well-known element of botanical Latin is the scientific name, formally known as a *binomial epithet*. This usually consists of a genus name in the nominative, capitalized by convention, and an adjectival species name that complements the genus name.[18] The concept of binomial nomenclature is often credited to Carl Linnaeus, but in fact, his work lay primarily in systematizing binomials and solidifying the processes of naming and describing plants. There was discussion about and deployment of the binomial format in earlier botanists, such as Kaspar Bauhin (1560–1624), Joseph Pitton de Tournefort (1656–1708), Augustus Quirinus Rivinus (1652–1723), and John Ray (1627–1705). The form also appears in the medieval herbal tradition with considerably less formality, although serving a similar function, distinguishing one species of a plant from another. Binomials can in fact be traced all the way back to ancient Rome. Pliny the Elder (*c.* 23/24–79 CE) made the transition from discussion of the animal world to that of flora in Book 12.1 of the first-century CE *Historia Naturalis* by claiming that, "the natures of all the animals which can be known reveal themselves thus, both in general and by reference to their parts (*generatim membratimque*)." Pliny was speaking of animals, but this shows that the concept of specifying and identifying a plant by naming its generic and specific characteristics was not new to the early European botanists. One of the more remarkable and consistently overlooked features of Pliny's text is his heavy use of binomials,[19] although he was inconsistent in his application. Linnaeus' definition of a binomial as a genus name accompanied by a species name[20] was novel in its specificity and insistence on adherence to its specifications, but the actual format was not new.

Pliny also discussed the confusion engendered by nomenclature that applies to more than one plant, and the opposite problem of plants with more than one name.[21] Similarly, in 1737 Linnaeus in his *Critica Botanica* stated that where there is one genus, there shall be one name (*unicum ubi genus, unicum erit nomen*).[22] This was the precursor to Linnaeus' description of his methodology, found in the *Ratio operis* of the *Genera Plantarum*, that "everything that can truly be distinguished by us depends on a clear method, with which we distinguish things that are alike from things that are unalike."[23] Linnaeus provided three very important clarifications of his method and project. First, he declared that plants (*vegetabilia*) are "known" (*nota*) to anyone who already knows how to join similar (*similia*) things to one another, and how to separate dissimilar things (*dissimilia*) from one another.[24] Linnaeus was promulgating a particular scientific approach that applies to the entire natural world: the individual trained to think about nature in this manner already has the tools to approach the study of plants. The raw material for accurately naming, categorizing, and describing a large number of plants is already available to the botanist: this material is not in need of discovery, but simply organization. This leads to the second clarification, quoted above, that the botanist knows how to assign names to plants that are intelligible to everyone. With this claim, Linnaeus was hinting at a system of nomenclature that is not *ad hoc*, but rather adheres to a prior system of classification: "similar plants with similar names" is a reference to the application of the same genus name to all species that fit the criteria of classification for that genus. Moreover, the criterion that these names be intelligible to all was plausible for Linnaeus and his contemporaries in eighteenth-century Western Europe. The idea that Latin names might not be clear and meaningful to future scholars was unthinkable and yet that was precisely the fate of botanical Latin in the twentieth century.

Linnaeus' third methodological clarification concerned the relationship between nomenclature and classification. He asserted that generic and specific names "should be certain and fixed, not variable, inconstant, or applicable to all manner of situations." Moreover, these names must be given to fixed genera, and if this rule is violated and if categories waver, "then their names also waver, and as a result, so too do the doctrines of the botanist."[25] Linnaeus was insisting on a 1:1 ratio between plant and name, which is not to be altered. In order to ensure this, it is also necessary that the genera themselves be fixed, categorized, and defined ahead of time. By "genus," Linnaeus was not referring simply to the taxonomic level between family and species, but to any fixed category used to diagnose a plant in order to distinguish it from others. This is what he was referencing when he declared that the very doctrines of botany would totter if the genera wavered. If we assume a 1:1 ratio between genus and name and then move the goalposts by changing the definition of that genus, along with what constitutes a genus within the system as a whole, then the ratio becomes 1+n:1 and there is no longer any consistency. The stability of botanical nomenclature was so important that Linnaeus cautioned his readers about the deleterious effects of its absence in the science of botany. The concept of a unique, fixed, and intelligible binomial epithet is also found in Pliny the Elder, for whom the issues at stake were knowledge of plants and descriptive names in Latin rather than the vernacular. Binomial epithets can therefore be traced back to antiquity, allowing us to ascertain and identify particular definitions and uses of the format.

Moreover, binomial nomenclature does not and did not ever exist in an intellectual vacuum. It was embedded in the broader concept of "botanical Latin." Establishing what constitutes a binomial epithet at particular stages in the history of botanical writing is relatively simple, but the corresponding issue of what constitutes botanical Latin at those same stages is far more complex. The slow transition from Pliny's prose to the modern set of highly formal rules and recommendations for composition is a continuum on which the transitions from point to point are more like fuzzy boundaries. Methodologically, it is therefore practical to focus on individual writers and the ways in which they use the Latin language to describe plants, the vocabulary they employ, the grammatical and syntactic constructions they prefer, and how they interact with the ancients and their own, contemporary intellectual milieu. Using these criteria, we can draw together the following list of the general features of modern botanical Latin, as seen from roughly Linnaeus' time to the abolition in 2011 of the 1935 Latin rule:

1. Binomial epithet, consisting of the genus name in the nominative and a species name (usually adjectival) that complements the genus name.

2. Diagnostic formula, consisting of paratactic ablative constructions that provide the unique features of the plant, in which punctuation, especially commas, play a crucial role in deciphering the ablative constructions.
3. Heavy use of prepositions + ablatives to describe geographical distribution.
4. Descriptions using the nominative and simple present active indicative verbs, in which a plant is compared and contrasted with other plants, following certain long-established orders and highly specific terminology. Punctuation and printing idiosyncrasies such as odd kerning are employed to draw attention to certain terms.

Some of these features appear in Pliny, some in early-sixteenth-century botanical works, and most in later botanists. The formal set of rules that constitute modern botanical Latin post-date the Renaissance, but the origins of a select few features of botanical Latin go back much further. The first half of this book will therefore cover the development of botanical Latin at two points on the historical continuum: Pliny the Elder's first-century CE text *Historia Naturalis* and the sixteenth-century Latin herbal of the German physician Otto Brunfels (*c.* 1488–1534), which I argue was significant to the development of botany and botanical Latin. The phrase "development of botanical Latin" refers to the stabilization of both binomial nomenclature and the rules and recommendations for the composition of descriptions and diagnoses. In spite of Brunfels' reluctance to position himself as an authority, his text was the first realistically illustrated herbal to be printed in Europe, placing it at the forefront of the development of botany as a scientific field. Moreover, Brunfels cited Pliny the Elder as his main inspiration and appropriated key elements of Pliny's work on plants, tying Brunfels both to antiquity and modern botany.

The original Latin found in Brunfels' herbal, long overlooked in favor of its realistic illustrations, functions as a proto-botanical Latin and can be positioned in contrast to its modern iteration. Despite this, it is difficult to pinpoint when botanical Latin began its transformation into its modern form. For instance, Stearn argued that botanical Latin is grammatically quite similar to classical Latin and that it primarily differs by virtue of its highly technical vocabulary. He designated 1650 as the approximate date when technical botanical vocabulary turned to specialization in ways that were unattested in antiquity.[26] Stearn also argued that it was with Linnaeus in the eighteenth century that this more specialized botanical vocabulary developed into "botanical Latin." It is true that *formal* botanical Latin, the partially and fully stabilized forms of the genre that populate the eighteenth through twentieth centuries, is a product of that era. The seeds of botanical Latin, however, go back much further, to the early sixteenth century and specifically to Otto Brunfels' much maligned herbal.

Methodology

There are two methodologies at work in this text. The first involves the approach to the subject matter. Botanical nomenclature is easy to define and its development and formalization throughout history is well-documented. The slow transition from Pliny's prose to the modern set of highly formal rules and recommendations for composition is, however, a much less straightforward path. Despite this, it is possible to focus on individual writers in the history of the study of plants and to consider the ways in which they use the Latin language to describe plants, the vocabulary they employ, the grammatical and syntactic constructions they prefer, and how they interact with the ancients and their own, contemporary intellectual milieu. The texts and authors discussed throughout this book have all been chosen as representative case studies in which modern botanical Latin's development can be identified. The second methodology involves the characterization of the historical changes in botany and botanical Latin. In his work on the Renaissance reception of Virgil, Kallendorf (2020) argues that the early printing history of his poetry in Rome was the catalyst for its enduring but sudden popularity.[27] Kallendorf employs transformation methodology, a Classics-oriented approach to the Renaissance reception of classical texts. In this model, transformations are reciprocal changes that occur between a reference sphere and a reception sphere. These changes can be attributed to agents (which can be events, people, or even technology) from the reception sphere, which is altered as a result of appropriating some feature from the reference sphere.[28] The form of these transformations is not conditional, or one-way, but an effect and a cause at the same time, involving reciprocity on the part of the two spheres, both reference and reception.[29]

In some ways, transformation methodology is focused less on individual people and more on processes, the agents of which may be individuals who engage with what the authors call the reception sphere. This sphere is akin to Foucault's episteme, which he defined as a series of discursive formations, patterns, and regularities that can be identified in a particular discourse, for example, pre-Renaissance natural histories that did not engage in classification.[30] It is also linked to Pamela Smith's use of the term "discourse" to describe what is meant by "science," which serves to highlight the literal meaning of the term *scientia* (what is known). A *discourse* is simply a set of facts that serve an intellectual purpose in a particular time and place, much like the Kuhnian concept of a scientific paradigm: what we consider to be knowledge is always subject to change and the limitations of our times.[31] A reception sphere is more specific than this, defined by its interactions with a reference sphere. The agent of change is a part of the domain of discourse, engaged with discursive formations, but they also alter the object of the discourse, which creates the reference. This kind of transformation is described as *allelopoiesis* (mutual creation), the reciprocal, biconditional nature of the process of reference and reception, of influencing and being influenced by one another. For its validity, *allelopoiesis* assumes and relies on both directions acting as agents simultaneously.[32] Prior to the development of transformation methodology, Eisenstein (1979) had also worked on the Renaissance reception of classical antiquity and concluded that the printing press was an agent of transformation because it altered the circumstances of reading. Books were printed, circulated, and utilized in fundamentally different ways, so that the *ars artificialiter scribendi* was also the art of reproduction and duplication.[33] In transformation methodology and *allelopoiesis*, that unidirectional understanding of the transformation process becomes productive and reciprocal, resulting in the creation of both the reference sphere and the reception sphere.[34] To an extent, we can see elements of this in Mikhail Bakhtin's dialogic principle, according to which there is a fundamental disunity between text and reader.[35] Dialogism assumes a text's polyphony, which Bakhtin applied to its "interpretation," rejecting the idea that truth is external to and independent of it. Rather, there is a "discourse" that can be identified, which consists of the interactions between the many voices and is characterized by a lack of harmony. Likewise, Foucault argued that texts show us what to do with them and that this self-descriptive feature is precisely the reason they need no interpretation. He attributed the disunity between text and reader to the discursive structure of the modern episteme, which is characterized by the presence of a "primary misunderstanding."[36] In the case of *allelopoiesis*, this disunity is applied to cultural reception that is based on textual transmission. The idea is not that an earlier culture exists or is literally constructed *post hoc*, but simply that our ability to reconstruct cultural mores and customs is limited by the artifacts to which we have access. Since these artifacts, especially textual artifacts, are rarely if ever transmitted intact and without contact with later agents, it is inevitable that they are altered, modifying in turn our conception of the artifact itself and its historical and cultural context.

Transformation theory is inherently interdisciplinary, interacting with numerous fields, in particular reception theory. It is also retroactive, looking back to the same degree that it looks forward and taking into consideration every element of the transformation process.[37] Applied to the reception of Pliny in Otto Brunfels' text and to the development of botany in the sixteenth century, all of the relevant texts work together as a collective agent in the transformation of Pliny's *Historia Naturalis.* This process creates a novel reference culture (Pliny's text) and a new reception culture (Brunfels' text), which in turn sets up the necessary circumstances for specific transformations of the science of botany and its expression in the Latin language.[38] In the model described by Kallendorf and Bergemann et al., there are subcategories of transformation, including reconstruction, the purpose of which is to reinstate that which has been either lost to history or only partially safeguarded. In this process, preservation of the ur-text is the only goal, apart from establishing authenticity. There is no exegetical element and no textual interpretation involved.[39] Similarly, the early printed editions of Pliny's *Historia Naturalis*, and the critical reactions to them, are reconstructions that purport to be authentic presentations of Pliny's original text, despite their reliance on the manuscript tradition and the inevitable scribal errors and alterations it entails. The manuscript tradition also relies on medieval florilegia, the subject-specific compilations of Pliny that put into digest form such topics as Plinian medicine and herbal remedies. This description of reconstructive transformation also provides a useful bridge between the issues of an episteme, a reception sphere, and a domain of discourse.

Specifically, however, Otto Brunfels' 1530 herbal is an example of a further transformation subcategory—supplementation—that is subsumed under reconstruction and includes the element of interpretation of the reference object. This is crucial to the argument that Brunfels' herbal is more important to the development of botanical Latin than previously thought, precisely because he did more than simply regurgitate the opinions of others: he inserted his own opinion whenever the cacophony of authoritative voices got too loud. Following Brunfels, there are innumerable other points on the continuum of botanical Latin, including Leonhard Fuchs' 1542 *De Historia Stirpium*; the seventeenth-century turn to technical diagnoses; the European fascination with "New World" plants; Linnaeus' work on classification; and the nineteenth-/twentieth-century distillation of the entire history of Latin botanical texts into botanical Latin. More recent points on the continuum include the institution and subsequent rejection of the 1935 Latin requirement and the current debates about Latin nomenclature, such as honorifics, the encoding of offensive, racist, and colonialist ideas and terms in plant names, and the deprivileging of indigenous names. Latin has already been replaced for diagnostic phrases. It may, eventually, be supplanted by other natural languages for nomenclature as well, at which point botanical Latin will become an historical artifact. Yet the philosophical issues of discoverability, knowability, and universal intelligibility remain. They are not tied to one natural language and cannot be resolved by switching to another natural language.

The changes in botany in the past few decades have had an unexpected result: the collapse of the ideal of universal intelligibility, which has long been assumed to be embedded in the Latin language. This book traces the origins of this ideal, along with other key features of botany, such as the premise that plant names, descriptions, and illustrations should all contribute to the plant's discoverability and knowability. In a sense, therefore, this book is also about the dissolution of botanical Latin. It begins by tracing its development and the pursuit and entrenchment of certain corresponding scientific ideals, only to highlight their collapse in the twenty-first century, all of which amounts to the "legacy" of botanical Latin, a term that does not refer to benefaction or heritage, but to repercussions, consequences, and aftermaths. It refers to the ways in which Latin developed into a modern tool for scientific communication and in particular how Latin continues to be used in those contexts. It involves the philosophical and ethical underpinnings of botanical Latin and critiques and interrogates the assumptions about knowledge of plants that are embedded in Latin scientific nomenclature and diagnoses. The methodology Critical Ancient World Studies (CAWS) critiques pretty much every aspect of the field that has been treated as objective and fundamental and "dismantles the structures of knowledge."[40] Botanical Latin may be scientific, but it has been formed by and subjected to the same knowledge structures as other genres. The current debates in the field of botany stem from a moral issue, not a scientific one. The resolution to these debates lies in de-privileging (not wholesale rejecting) the names Europeans have foisted on indigenous plants. It also entails reconsidering the implicit assumption of the scientific ideal of universal intelligibility. Critique of this ideal does not entail throwing out all of the knowledge we have gained about plants, such as taxonomic ranking, propagation, habitats, and DNA sequencing, but simply rejecting the assumption that to know a plant is only to know it in these extremely limited ways.

1

Pliny the Elder's *Historia Naturalis*: The Origins of Botanical Latin and Nomenclature

Gaius Plinius Secundus (*c.* 23/24–79 CE), better known as Pliny the Elder, was the most prominent natural historian of first-century Rome. His 37-volume, encyclopedic *Historia Naturalis,* dated to roughly 75–9 CE,[1] was a compilation of knowledge about the entire natural world, including geology, minerals and mining, agriculture, astronomy, zoology, and botany, culled from ancient sources, contemporary second-hand accounts, and his own investigations. Books 23 through 29 are considered botanical, describing the medicinal uses of plants and the state of the medical profession in Pliny's time. These seven books are indicative of both Pliny's personal style and a broader, Roman "scientific" style of description that would, by the twentieth century, become encoded in what is now known as botanical Latin. Between Pliny's text and the twentieth century, however, is a crucial point on the continuum of the development of botanical Latin: the German physician Otto Brunfels' 1530 *Herbarum Vivae Eicones* (Living Images of Herbs). The features of the *Historia Naturalis* that deal with plant life are reference points for Brunfels' herbal, which in turn played a role in the development of botany in the sixteenth century.[2] Every botanical element of Pliny's text is an agent in the transformation of the Latin used by Romans to describe plants, into modern, formal botanical Latin. This is not a direct, causal relationship, but rather, in accordance with transformation methodology, Pliny's text serves as a reference sphere for the early botanical texts. It is also a reception sphere, in which it is possible to locate both how Pliny reacts to his botanical predecessors and how he transforms what he has inherited from them. Pliny made much of the fact that he was the first to write down what he knew about plants in the Latin language. It was important to him to establish Rome and its native tongue as the locus of and a viable communicative tool for the study of nature. The reasons for Pliny's insistence on the importance of Latin are, of course, different from the reasons why Latin was used by European scientists, but they both assume the ideal of universal intelligibility. A second point of reference is that in his rhetorical digression in Book 25, Pliny emphasized the importance of having a more precise, systematic process for the naming and discovery of plants. Fundamental to this discussion is his concern for "lost" plant knowledge.[3] This deficit, he argued, was due to imprecision in naming conventions, with the result that it was difficult to know if one plant name referred to the same plant as another plant name. The overall loss of plant knowledge was also the result of a tradition of secrecy surrounding plant lore, as well as inferior methods of transmitting knowledge about plants. Pliny was engaged in a project that is twofold: to recover lost knowledge of plants and to ensure that once regained, this knowledge was properly preserved and passed on to posterity.

Within the framework of transformation methodology, Pliny's text itself is a reconstruction of other ancient writers, an *allelopoietic* transformation of the earlier natural history sources with which he interacted and a reference sphere in and of itself. Specifically, in Books 23 through 29, Pliny appropriated the work of Theophrastus of Eresos, engaging in a rhetorical digression in which he critiqued the methodology of Asclepiades of Bythinia, the Greek rhetor-turned-physician, rejecting features of the Greek tradition and retaining others, making them distinctly Roman. Although there is much that is original in Pliny, there is much that has been borrowed from his predecessors, particularly in the sections of his work that deal with plants and medicinal herbs. In a cycle of reference > reception > reference, one cannot treat Pliny's text as the reference sphere for Brunfels' Latinity and style without considering the ways in which the *Historia Naturalis* is itself a reception sphere that interacts with and transforms its own reference sphere. These reactive and *allelopoietic* features of Pliny were also appropriated by Brunfels and incorporated into botany and botanical Latin, including his use of the binomial format, his concern for correct naming and identification of plants, his rejection of Greek methodology in favor of experience, and the ethics of the study of plants.

Scholarship on Pliny

Pliny's status as a natural history authority, especially in the field of botany, had declined significantly by the end of the Renaissance. The early-medieval reception of Pliny was marked by the production of florilegia, extracts, compilations, and additions to the *Historia Naturalis*.[4] The demand for these kinds of interactions with Pliny can be explained in part by the inaccessibility of his monumental work, with its encyclopedic length and structure,[5] but it also indicates a generally positive view of Pliny as a natural history authority. By the twentieth century, the *Historia Naturalis* was seen as a mostly unoriginal, lengthy compilation of earlier Greek and Latin naturalists.[6] Critique of Pliny's Latinity can be traced back at least to Eduard Norden's 1898 declaration that his stylistics are "dreadful,"[7] an opinion supported a century later by Goodyear's 1982 summary of the author as a "pedant" incapable of coherence.[8] Garrison (1931) also betrayed a less than effusive opinion of both Pliny and the value of ancient approaches to natural history, describing the *Historia* as the substitution of old wives' tales about flora and fauna for the more scientific observations of the naturalists who preceded him.[9] All of these can be contrasted with Foucault's remarks on the development of Renaissance natural histories, in which the archaic signification of the word *historian* as "one who sees," rather than one who retells what they have collected and compiled, was reclaimed.[10] Foucault was referring to the original meaning of the Greek term *historia* as it was used by Herodotus, for instance, as an "inquiry" or "investigation." By the time of Aristotle and his disciple Theophrastus, it had evolved to mean something akin to the systematic process of obtaining knowledge or information.[11] In Pliny's time, the term already had a proto-scientific use, rendering his work more than a narrative regurgitation of prior sources. The *Historia Naturalis* also consists of original, firsthand observations by Pliny himself and where he did summarize the ideas and work of others, he added in commentary that provides useful insight into the use, reception, translation, and dispersion of natural historical texts in antiquity. Moreover, Pliny's rhetorical digressions are rich with crucial information about how his predecessors reasoned through their contact with the natural world. These digressions, together with Pliny's penchant for moralizing, irritated Wallace-Hadrill (1990), who pointed to such excursions as an unscientific weakness.[12] Yet without them, or those of other writers such as Galen, we would not know, for example, that the Greek tradition of reasoning from first causes about medicine and herbs, which Asclepiades employed, was deplored by later physicians.[13]

Morton (1986) marks a turn toward a more positive take on Pliny's project, pointing out that there is a substantial amount of material in the *Historia Naturalis* that cannot be traced back to any particular author, an indication that Pliny himself is the likely source.[14] Wallace-Hadrill (1990) argues, despite his other reservations about Pliny, that the *Historia* is coherent and that Pliny's rhetoric helps to place its subject matter (Nature) in context for a more complete picture. Adams (2005) notes the parallels between Pliny's technical writing and Roman poetics, which share syntactical features because they both have a need for economy of expression.[15] Doody (2010) calls the later antique tradition of florilegia a "dismantling" of Pliny, who was more than just a pedant regurgitating the ancients. The *Historia* stitches together disparate elements and sources to create a highly original and authoritative work.[16] Doody and Jackson (1988)[17] agree that the completeness and breadth of the *Historia* was its primary weakness and that the dismantling of the text into smaller, more digestible, specialist texts, primarily geographical, medicinal, and herbal, rendered it not only more accessible, but less prone to error, simply by virtue of their shorter length. Nutton (2012) focuses on the importance of Pliny's use of the Latin language for the preservation of Greek medicinal knowledge and points out that by the end of the third century BCE, there was already a technical Latin medical vocabulary from which Pliny was able to draw.[18]

The *Historia Naturalis* is a rich source of information for anyone interested in the methods, materials, and philosophical foundations of doing medicine, botany, or simply observation of the natural world in antiquity. Books 23 through 30 of the *Historia Naturalis* show how Pliny transformed these processes as they were done by his predecessors and produced something new. The Romanization of the works of the Greek writers on whom Pliny relied was not just a process of translation from one language to another. It also involved the pointed rejection of the Aristotelian methodology employed by both Theophrastus and Asclepiades. Pliny cast aside pure deduction and arguments from first causes in favor of experience and the wisdom of folklore. In addition, he believed that Nature had produced a number of plants specifically for medicinal purposes, for the benefit of humankind. Any approach to the study of these plants that is exclusive or makes herbal remedies inaccessible to people must therefore be rejected. This is why Pliny focused on ease of identification, the sharing of knowledge, and expertise gained in the field as opposed to book-learning. These three features of Pliny's text were appropriated by Otto Brunfels in 1530 and given new meaning, transformed into a Renaissance phenomenon that was tied to and embedded in the development of botany as a scientific discipline and to the burgeoning of more and more formal methods of expression in the Latin language.

Binomials and Plant Identification in Pliny

Binomial nomenclature is the method of naming biological entities after their genus and species. The format consists of two plant names, the first of which indicates the genus and the second of which is the species name, usually an adjective that distinguishes the plant from others. Despite its association with the eighteenth-century botanist Carl Linnaeus, the use of this format can be traced to Pliny's *Historia Naturalis*. Although he was inconsistent in his use of binomials, he often used two names when he knew of more than one kind of plant, generally using, for example, *maior* (larger) and *minor* (smaller) for species names, or *nigra* versus *alba*, or the lands in which they originated. Pliny's adoption of these binomial forms is due in part to the fact that he was concerned with precisely and helpfully naming plants. This meant employing accurate names that have identifying information embedded in them and are unique to one particular plant. Similarly, he was aware of the fact that the same plant may have different names depending on the place. For instance, he claimed that the oldest plant to have been discovered was the peony (*paeonia*), which still had the name of its finder, the physician-deity Paeon. It was also called alternately *pentorobon* and *glycyside*, which created a further difficulty, since "the same plant may be called one thing in one place, and something else in another."[19] Two of the same nomenclatural difficulties that propagated Linnaeus' system of binomial epithets were common to Pliny's era as well: using the same name for different plants (or failing to distinguish between different species of the same genus), and using different names for the same plant.

Rather than engage in either a detailed mapping of plant names to one another, or some system of ranking by which preferred names might be selected, Pliny focused his attention on the problem of signification: plant names should be in some way descriptive and should not indicate their discoverer. In support of his campaign for more informative plant names, he critiqued the treatment of plants by the ancients as being too wondrous and full of marvel. Viewed panoramically, Nature is awe-inspiring, but if one takes the time to focus on the details, the whole becomes more approachable. In the case of medicine, the accumulation of knowledge about the minutiae of the plant world, such as leaf shape, flowers, habitat, and medicinal uses lessens one's credulousness. Knowledge gained from experience, at a granular level, renders one's contact with the larger world more informed and aware. Pliny's criticism of viewing nature as *mirata* (wondrous) was rooted in his belief that long years of close contact with and experience of nature's productions generate grim, hardened utility and common sense, not childlike wonder. Experience and attention to detail provide an individual with the necessary tools to begin to approach Nature on a broader scale. This critique of Pliny's provides us with criteria with which to examine his own writings: the presence of binomials whenever a plant has more than one known species or variety; attention to identifying characteristics and their accompanying vocabulary; and an effort to refrain from expressions of wonder. Binomials are just one piece of the puzzle, the aim of which is to be detailed, informative, and precise when writing about plants.

Imprecision in nomenclature, incorrect identification, and overall gullibility with respect to Nature all played a part in the general lack of *cognitio* (knowledge) of plants in ancient Rome. At Book 25.6, Pliny described how there were two reasons that plant knowledge had devolved in his time. First, herbs were considered the domain of illiterate country folk, who alone live their lives among plants and therefore have experience of them. Pliny implied that wealthy aristocrats had not concerned themselves with the exigencies of plant use cultivation because they deemed it beneath them. The second reason for the devolution of plant knowledge was that there were, "for many known plants, missing names" (*multis etiam inventis desunt nomina*). As an example, he referred to an unnamed plant that, if buried in the edges of fields, would keep birds away. Though he was careful not to state any such thing outright, Pliny implied that, because rural folk have a habit of using plants without bothering to assign formal names to them, they could use some helpful input from educated individuals such as himself. Yet it seems unlikely that a plant that is *inventa* (discovered, or known) does not have some kind of name attached to it. Within a particular community, a singular name in the vernacular could easily refer to exactly one plant that everyone in the community associates with that name. The "missing names" to which Pliny was referring were in fact full and descriptive; they just did not have the form that he recommended.[20] Pliny was advocating for Latin names of a specific format, because he thought that this was the most universal and understandable by everyone in the empire. But Pliny was sidestepping the deeper issue, the connection between the "discovery" of a plant and its naming. It is unclear what Pliny meant by *invenio* (to discover, or know) and *inventa* (known) in this context, if he was advocating for a proto-scientific system of identification and assigning of names, or implying that a plant that had been used for centuries by peasants could not really be said to be "discovered" if it did not have the kind of name of which Pliny approved. Throughout the *Historia Naturalis*, Pliny often provided the Greek name for some plant or animal, while acknowledging that there was no name for it in Latin,[21] and even described a plant whose name was *Anonymus*, because it "has found a name by not finding one."[22] Pliny claimed that this plant was indeed well known in Scythia and used by the Greek physicians Hicesius and Aristogiton. Thus, despite apparently having no name, it was in fact *inventa* (known).[23]

One of the underlying reasons for Pliny's concern about missing plant names involves the naming conventions of ancient Rome, specifically, the transition from a binomial system to the better known *tria nomina* (three names). Roman citizens had a *praenomen* (first name), *nomen* (clan name), and a *cognomen* (last name, which often was a nickname). The same few *praenomina* were recycled, with almost everyone being Marcus, Lucius, Gnaeus, Gaius, and so on. The *nomen* was the *gens* (clan) name and therefore provided more descriptive information about the individual. Pliny sometimes, though not always, used the term *nomen* like *praenomen*, as in the phrase *desunt nomina* ("their names are missing"). Here, he likely meant *nomen* either as the *gens* name or possibly both a *praenomen* and a *nomen*, that is, somewhere in between the two in terms of specificity. The most granular of the *tria nomina*, though, was the *cognomen* and Pliny's use of this term is not interchangeable with the other two. Some individuals, Pliny said, having made an important botanical discovery, have been rewarded and made more illustrious by having their *cognomen* attached to the plant and to the memory of humankind.[24] The Roman *praenomina* were too few in number and therefore too non-specific to be used in such a manner. *Nomina gentilicium* (clan names), too, were not specific enough, since they could be shared by any number of citizens from the same *gens*. It is therefore the *cognomen* that is used to name a plant after its discoverer, since it functioned in much the same way as the species names in binomial epithets. They provide information about the individual's characteristics, such as their personality, physical form, facial features, occupation, place of origin, and their successes.[25] Another reason why the binomial format shows up in Pliny and not in the Greek writers is that the Latin language is less prone to synthesis than Greek. Nybakken (1959) explains how Greek in particular is ideal for the construction of "economical, descriptive, and euphonious compound terms,"[26] which, by virtue of being composed of more than one word, have very complex meanings embedded in them. Because Latin superseded Greek in botanical nomenclature, most examples of this feature in Greek are found in medical or pharmaceutical terminology, such osteoporosis, myopathy, and encephalopathy. Latin simply does not allow for such variety or originality and, as a result, by insisting on using Latin, Pliny had to resort to the binomial format in order to express the same level of detail.

The convention of assigning the name of the discoverer to a plant, whether the *cognomen* or *nomen*, is the easier of the two nomenclatural difficulties to handle. One simply needs to instate a new protocol that involves assigning meaningful names to all plants. The larger issue is that this is of no use if the proponents of the field close ranks. In Pliny's opinion, the worst of the reasons for lack of general plant knowledge in Rome was that those who did know such things did not bother to teach them to others.[27] As an example, Pliny cited *cynorrhodon*, which cures the hydrophobia that results from a dog bite, but which most people did not know of, because the authorities did not write about it. Its medicinal use was only popularized when a soldier who had been bitten by a dog received a letter from his mother informing him of the cure, about which she had dreamed. In addition, Pliny was disturbed by the fact that no one had ever bothered to devise a systematic method for the identification and collection of plants. Despite his advocacy for the knowledge of plants gained from rural inhabitants, he thought that there was a need for a methodical survey of this knowledge and the assignment of useful, descriptive names to all plants. Pliny complained that all previous works on plants and herbs were nothing more than *ad hoc* compilations of lore surrounding them. However, this is not entirely true, or at least, it is not the most accurate description of the works of Greeks such as Theophrastus. Theophrastus did not employ a systematic method of identification and collection, but his *De Causis Plantarum* and *Historia Plantarum* are highly organized and use the Aristotelian method of deduction. Pliny is more likely to have been criticizing the Roman authorities who came before him.

Nomina nuda (Bare Names)

Beyond nomenclature, there was another issue with the promulgation of knowledge of medicinal herbs. Greek writers[28] who dealt with plants, such as Crateuas, Dionysius, and Metrodorus, employed a methodology that Pliny viewed as peculiar: they would provide an image (*effigies*) of a plant, under which they would include its properties.[29] This had the potential for inexactness since "a painting is misleading in its variety of colors, especially in the emulation of nature."[30] There was also the risk of introducing mistakes when copying illustrations, an indication that the twin issues of manuscript variants and scribal error were on his radar. Moreover, in these illustrations, only one image would be provided of a plant, which can have quite varied appearances depending on where it is in its growth cycle. In response to these exigencies, Pliny noted, other Greek authors chose to forego illustrations entirely, even though they can make identification in nature easier, using words alone to describe plants. In some cases, these individuals even neglected to provide a full name for the plant, preferring to describe a plant's properties and medicinal uses and to use "simple" or "bare" names (*nudis plerumque nominibus defuncti*), as opposed to more specific ones. For Pliny, it was not enough to describe the potency or strength of a plant, since in order to avail oneself of a plant's uses, one first has to find it. In this custom, if an individual wishes to learn the medicinal properties of a plant, they need to consult someone who already knows how to find the plant. By contrast, had this information been provided in a more systematic way, accompanied by illustrations and names that actually provide insight into the plant's properties, that individual would be more independent and able to pass this information on to others.[31] The problem is that these "bare names" are in the same general category as the "missing names" above: they are only bare to those who do not understand them. In fact, Pliny claimed that this kind of knowledge about plants, despite having been handed down in less-than-ideal ways, by word of mouth and without the benefit of illustration, is actually easy to come by (*nec est difficilis cognitio*). But if knowledge of herbal medicine is so readily available, why does it matter so much that the Greeks handed it down by word of mouth, with "bare names" and without illustrations? Pliny provided two explanations. First is the intentional obfuscation of his predecessors in the study of plants. Those who use "bare names," or name plants after themselves, or employ inferior methods of knowledge transmission are not simply guilty of a sin by omission. They are purposefully keeping knowledge of plants to themselves. For Pliny, and later on for Otto Brunfels, concern about accurate naming and transmission of knowledge was still a moral issue. By Linnaeus' time, however, the *innovatio* (variety) of plant names, while still a major problem that warranted the publication of the *Fundamenta Botanica* in 1736, had transformed into a scientific problem. Second, Pliny was hinting at a larger claim that he would make later on, namely, that book learning cannot replace practical experience when it comes to plant lore, whether it be propagation, cultivation, or use in medicine. Pliny did not use *verba* (words) or *scripta* (writings) but rather *sermo* (conversation) to describe this method of knowledge transmission. *Sermo* as a method of transferring knowledge, a reference to schoolhouse lectures, was insufficient because of the inevitable corruption of the original statement once repeated. Pliny seemed to think that there are plant names that are the opposite of *nudus*, that are instead full, laden, clad with information, and serve the purpose of and allow for accurate identification. The Greek writers did not think such names were necessary.[32] Pliny claimed that there is no need for glorious quests in order to gain knowledge of plants: one need only examine them in person.[33] He therefore rejected book learning in favor of knowledge gained from experience.

Pliny's recommendations for plant nomenclature can be extended to the field of medicine, since identification and accurate, readily available information were his main concerns and plants were used medicinally. We can see how he applied these same criteria to the names of diseases. In Book 26, Pliny discussed a number of new, disfiguring facial diseases that were popping up all over Rome, but which were unknown to the ancients. He provided a lengthy account of the symptoms, transmission, effects, and treatment of these skin diseases and marveled that some of them, such as leprosy, would pop up in Italy and then disappear. For other diseases, even their name became obsolete, as for example the disease called *gemursa* by the ancients,[34] which affects the toes, but whose name was now completely forgotten. There was no contemporary Latin equivalent for this term, because the disease had disappeared and was therefore no longer in need of a name. He then turned to the disease *colum*, an intestinal sickness that had not disappeared but had become endemic, having arrived in Italy during Tiberius' rule. Because the emperor was the first to suffer from it, the people of Rome had not heard the name until it was read out in an edict in which Tiberius cited it as an excuse for a sick day. In contrast to the disease *gemursa*, which Pliny expected his readers to recognize even if it was no longer endemic in Italy, the people had no frame of reference to *colum* and no "thing signified" they could identify. Although Pliny was discussing diseases and not plants, he was engaged in an astute discussion of the difficulties inherent in and the importance of naming and identifying things. As he asked with regard to this new disease name, *colum*, was it really necessary to invent a new kind of disease? Were the kinds that were already established too few, such that new ones were needed to instill fear? This conundrum is further complicated by the fact that humankind has brought the same number of problems on itself. Whereas Nature produces a number of works (*opera*) for our benefit,[35] mankind instead creates "work for itself through its own labor" (*sibi opera sua negotia*). Nature does not produce anything to its own benefit or detriment.

Pliny's remarks cannot be taken as anything more than the developing kernel of a natural system of philosophy, but he was aware of the fact that the plethora of issues humankind faces are in many cases their own fault. This includes the confusion surrounding naming conventions. What, indeed, is the difference between *colum* and other such diseases, that it needed its own new name? Similarly, the names of plants, if mixed up, resulting in the wrong signifiers, can create more work for humans than benefits. This is partially why Pliny so frequently explained the etymology of a plant name, because when the physical description of the plant is not sufficient to distinguish it from others, it may help to know why it was so named. This habit of Pliny's would not be necessary if plants did not have "bare names" and a person could divine key features from their nomenclature, and if the descriptions, whether visual and illustrative or verbal and textual, were accurate and precise.

Pliny's Rhetorical Digression

The beginning paragraphs of Book 25 mark the beginning of a rhetorical digression in which Pliny discussed certain concepts that are key to the argument that his text is both a reference sphere for early Renaissance botanical texts and a reception sphere in which earlier botanical sources are transformed. In the first two sentences, Pliny asserted that the ancients knew almost everything there was to know of plant lore, none of which they kept to themselves and all of which they passed on to posterity. "For nothing was untested and untried for them and in fact, nothing was secret, nor was there anything that they did not want to be of benefit to posterity."[36] Pliny contrasted this with his own time period, when ancient knowledge was kept secret and hidden away. By engaging in this secretiveness and suppressing the things elaborated upon by the ancients, Pliny's peers defrauded others of life itself. As a result, the chief aim of intellectual study in Pliny's time was to keep to oneself one's discoveries and in so doing, to forget what was once known to all, so that "inside each one of them the deeds of the ancients might perish entirely." In the study of plants, the private retention of one's knowledge has the negative effect of preventing others from knowing how to heal themselves and their loved ones with herbs. This tendency can be seen in the corpus of Greek magical papyri, in which most of the recipes involve incantations recited over plants and other mystical practices surrounding their harvesting and preparation. One example, however, stated that temple scribes had purposefully used false names such as Blood of Hephaistos (wormwood; or, *Artemisia absinthium*), because the populace could not be trusted to practice magic responsibly.[37] In the pseudo-Galen *Peri Antemballomenon* (Concerning Substitutions), the author discussed *pharmaka* (medicaments) that were used or spoken about in place of others. The author wished to reveal their true nature and names so that his fellow physicians would not accidentally cause harm to their patients.[38] Expressing similar concerns, Pliny asserted that this kind of secretive knowledge was not really knowledge because it was not shared. This revelation fleshes out the connection between discovery and naming. Pliny did not outright say so, but it was implied that the term *inventa* (known) could only be used of a plant if it was both known to a broad audience capable of identifying and using it and had a name that assisted this process.

Pliny also lamented the fact that his fellow Romans, despite their interest in anything utilitarian, seemed to have ignored the study of medicine.[39] All of this ties in with his claim that there are no best practices or mores in the discovery of new things. As a result, individuals kept what they found to themselves and the information that the ancients once knew and passed down had died off in Pliny's time. The custom of applying one's *cognomen* to a particular plant was also restricted to plants with medicinal uses.[40] If a herb has strengths associated with it, not simply pleasures, then not only does it bring glory to its discoverer, it is worthy of being passed down.[41] This is a far more important issue than it might at first seem. Romans who valued knowledge, utility, and strength seemed not to care for the study of plants. Moreover, when they did know of some product of Nature, they slapped their own name onto it and kept it private.[42] In contrast, Pliny claimed that it was a Roman freedman who knew the most about medicinal herbs.[43] At 25.3, Pliny explicitly named Pompeius Lenaeus, freedman of Pompeius Magnus, who had defeated Mithridates VI Eupator, the famed Persian-Greek king. Mithridates had learned that by ingesting poison one could develop immunity to it and had discovered a number of antidotes, including the one that retained his name (*primo inventa genera antidoti ex quibus unum etiam nomen eius retinet*).[44] Here, Pliny used the term *nomen* as opposed to *cognomen*, possibly because Mithridates was not a Roman citizen and therefore did not have the *tria nomina*, in which case, his "family name" was sufficient to identify him. Pliny was highlighting the fact that among the Romans, it was a non-elite person, a former slave, who had the most knowledge of medicinal herbs. He implied that this knowledge might have been learned from contact with the Persian-Greek king, who himself corresponded with the Greek physician Asclepiades.[45] Pliny claimed that Mithridates solicited Asclepiades, who sent his *Precepts* (*praecepta*) from Rome in his place, hinting that Asclepiades wrote about medicinal herbs and that these *praecepta* informed Mithridates' vast knowledge of medicine.[46] Mithridates was said to have had a large library of treatises on every subject and a large number of specimens, which eventually made its way to Pompeius Magnus, who then ordered Lenaeus, a man skilled in grammar, to translate them into Latin. According to Pliny, then, the knowledge of medicinal plants that the Romans possessed originated in a translation from the Greek. Asclepiades lived in Rome, but had not written his *Precepts* in Latin, nor had any of his colleagues bothered to translate them while he was there. Rather, it took a Roman to obtain the volumes, which were in the collection of a Persian-Greek king, to have them translated. Pliny eventually made clear why he was so concerned with this issue of translating Greek texts into Latin. In part, the transmission of knowledge has a moral component, which is the reason why in Book 29, Pliny made an argument for writing about Nature in original Latin.

Pliny also implied that the Romans who had contact with Asclepiades while he was in Rome were either uninterested in his writings on this particular subject, or there was a language divide among the Roman intelligentsia such that this knowledge was inaccessible to those who could not read Greek. In addition, Lenaeus is described as skilled in grammar, not in oratory, an interesting claim in light of Cicero's remarks on translating from Greek into Latin. In reference to the distinction between translating the sense of a passage or work (as an orator), and translating literally (as an interpreter), Cicero explicitly stated that he had translated Aeschines and Demosthenes, both Greek orators, not word for word, like an interpreter, but as an orator himself, "with the same ideas and the same forms of the ideas, or rather, the same figures, in words that are suited to our own way of speaking."[47] Pliny did not give any information about the nature of Lenaeus' translations, but it is worth considering that even in Pliny's time, there were scholarly discussions about how to capture the sense of an author's words when moving from one language to another. In the case of medical precepts, the importance of accuracy has a different tenor than it does for Cicero the orator. For him, the main concern was the argument and the fact that the entire point of an oration is to persuade. For someone involved in medicine, the stakes are considerably higher, since inaccuracy can result in sickness or even death. The so-called *ad verbum* style of translation that Cicero rejected was, however, suited for technical texts.[48] Pliny did not outright address this particular issue, but we can read between the lines and see that by discussing this minor historical occurrence, if indeed it even happened, Pliny was providing us with invaluable information about translation between Greek and Latin in the first century, about the status of medicine versus oratory, about the transmission of knowledge over time, and about *cognitio* (knowledge) of plants.

Although Wallace-Hadrill (1990) complained about Pliny's rhetorical asides, dismissing them as unscientific, the morals embedded in them are invaluable for deciphering the state of medicine in Pliny's time. Rizzo (1973) has pointed out that the words we use are an indication of what we value, that what we engage in dialogue about is a reflection of our beliefs, concerns, and values.[49] Pliny was not wasting his words by moralizing about handing down one's *cognitio* about herbs or by discussing the translation of Asclepiades into Latin by a freedman. The former was a reaction to the unfortunate custom among his contemporaries to hoard one's knowledge and the latter provided an example in support of his recommendation to reject elitism regarding authoritative sources. These, together with his remarks on the need for a system of discovery and meaningful nomenclature, do more than simply set the stage for his upcoming critique of Greek methodology. They serve as evidence for the argument that Pliny's text was itself a reception sphere, in which he did far more than simply reiterate the ideas of his predecessors. Pliny was providing his readers substantive suggestions for a method of doing "botanical" work that was both moral and resulted in accurate and reliable information.

Critiquing the Greeks: Asclepiades

In Book 26, in the middle of a discussion about an unknown and unnamed disease from which the emperor Tiberius claimed to suffer, Pliny suddenly switched to an overview of the ancient authorities who wrote about Nature's plant remedies. These, he said, were known for a long time, and long in vogue.[50] For example, there are volumes of Hippocrates, who first produced "precepts for the art of healing," that are full of mentions of herbs. But over time, actual experience, the most efficacious of teachers, especially in the art of medicine, gave way to *verba et garrulitas* ("words and chatter").[51] It was more comfortable to sit in a schoolroom engaged in listening to an instructor than to go into the wilderness and seek out various herbs on certain days of the year.[52] For this reason, students of medicine preferred to learn about herbs through books rather than through fieldwork, the main recommendation for which is proper identification, especially if it is assumed that the individuals in question did not have reference images. The importance of harvesting the correct herb in the proper growth stage, with the correct appendages, did not take precedence over physical comfort.[53] Despite this devolution in the quality and character of the art of medicine, nothing changed until Asclepiades, the master of rhetoric, abandoned his field of expertise for medicine. This was a crucial turning point because Asclepiades was sharp enough of mind to reject all of the medicinal knowledge of the ancients and withdraw the entirety of medicine into a system of causation, despite having no practical knowledge of herbs or remedies associated with them. By doing so, he made medicine into a matter of conjecture.

Asclepiades had identified five general principles for health and the treatment of ills: abstention from food; abstention from wine; massage; walking; and short trips to promenade around.[54] Pliny was neither derisive nor supportive of these principles, because they require no equipment, money, or expensive treatments. They are so common as to qualify as banal and even pointless.[55] However, regarding their effects on the people of Rome, Pliny turned to overt derision, calling Asclepiades' preferred remedies of wine and cold water an "empty artifice," by which he was accustomed to draw the minds of men to him.[56] Asclepiades was not even original, because Herophilus[57] had already investigated the causes of diseases, and Cleophantus had recommended wine as a remedy long before. Hoping to make a name for himself, Asclepiades preferred to be called the "giver of cold [water]," as Marcus Varro claimed (*ipse cognominari se frigida danda praeferens*).[58] It was apparently insufficient for the famed Asclepiades to be called *medicus* like other physicians. He needed a more distinctive epithet to ensure the endurance of his legacy. Other Asclepiadian remedies Pliny called *blandimenta*, since they were based on the simple pleasures in life, such as bathing, sleeping, eating, and drinking. So obvious were these remedies that were it not for a few "miracles" performed by Asclepiades,[59] Pliny was sure that his fame would not have endured. He was only angered by the fact that a man "from a most superficial group," who was inspired by the desire to make more money for himself with no resources, had suddenly dispensed rules for health to his fellow humans. The group in question included rhetoricians, whom Pliny considered vainglorious compared with trained physicians who save lives. Pliny's treatment of Asclepiades involved begrudging approbation of certain remedies that he promoted, such as steam baths, but he disapproved of the way he came to his conclusions. Asclepiades did not have the requisite training and as a result, his precepts were to be viewed askance.

Pliny focused primarily on Asclepiades' lack of experience, but his association with the Methodists and their refusal to draw any causal connections between disease and symptom was also an issue for other physicians. Galen referenced Asclepiades numerous times in his *On the Natural Faculties* (*ONF*),[60] in which he sarcastically noted that certain of his (and Erisistratus') theories were due to the heights of wisdom they had reached. Galen provided an in-depth explanation of Asclepiades' theory of kidney stones, followed by his own derisive take on it:

> It is my belief that Asclepiades never observed a stone that had been passed by one of those suffering from this ailment, or that a sharp pain preceded it in the space between the kidneys and bladder as the stone passed through the ureter, or that, once the stone had passed, the symptoms of both the pain and the urine retention immediately ceased.[61]

Galen therefore had the same general criticism of Asclepiades as Pliny, namely, that he was inexperienced and tried to explain physical ailments by reference to theories rather than experience. Had he ever actually treated patients and paid attention to their symptoms, he would have made the connection between abdominal pain in a certain area, the cessation of that pain, and the passing of kidney stones. Galen also explained Asclepiades' theory of how urine is formed in the bladder from the liquids we ingest, which he saw as blatantly incorrect, concluding that, "if he had ever studied anatomy, he would have immediately known." This conditional makes it clear that knowledge of biology gained through experience and not deduced from first principles is not only preferred, but actually makes for a better doctor. The *iatros* (doctor) who can marry experience of cause and effect, as in the relationship between a kidney stone's passing and the kinds of abdominal pain that precede it, with theoretical knowledge deserves the title of "Healer." For Galen, the moniker "Asclepiades the Physician" was both tongue-in-cheek and undeserved. It underscored the fact that he had no real claim to such a title, while also exposing the affectation of the appellation that both Pliny and Varro claimed he wanted: "Asclepiades the Wine-Giver," or "Asclepiades the Water-Giver," is almost comically accurate—he may have provided remedies, but he was no true physician.

Galen also made known his opinion of Asclepiades' methods at *ONF* 1.xiii.42, where he said that Asclepiades was so married to the elements on which he based his theories that he could not admit the truth about drugs. "He believed that agreement with the truth of such things was abandonment of the proofs that he had assumed."[62] Galen had already explained that Asclepiades thought, in contrast to Hippocrates and other real, practicing physicians, that drugs produced bile, phlegm, and other fluids. "Nor would he agree with these individuals, but instead, he claimed that each kind of [fluid] evacuated [from the body] originated in the drugs themselves."[63] Galen's response to this was to ask, both rhetorically and seriously, whether he should consider Asclepiades insane or simply inexperienced in the art of medicine. He explained how Asclepiades was only concerned with valid argumentation and with following his first principles to their logical conclusions. Asclepiades' writings were clearly derived from their logical principles, but they were also at variance with reality:

> If someone were to agree carefully with Asclepiades' writings, he would learn the logical consequence of such teachings, with their first principles, but he would also learn how they conflict with sensory experience.[64]

This was in contrast to Epicurus, who tried so hard to verify his experiences that he attempted to make them adhere to principle. Asclepiades was so concerned with preserving his principles that he ignored facts and would defend "whatever is logically consequent to his first principles" and so had no interest in any of his experiences.[65] Fortunately, Galen cited Menodotus the Empiricist, who had thoroughly refuted Asclepiades and demonstrated that his tenets not only did not correspond to reality, but even contradicted each other.[66] We can therefore identify at least three ancient authorities who questioned Asclepiades' motives, methods, and preconceptions: Pliny, Galen, and Menodotus. If Pliny is to be believed, the Roman people themselves eventually caught on to Asclepiades' lack of experience, but it is unclear if this preceded or followed the translation of his *Precepts* into Latin. It is possible that he was practicing medicine of some sort in Rome, but once his tenets were available in a language other than Greek, their variance with common sense and traditional medicine precipitated the decline in his popularity.

Critiquing the Greeks: Theophrastus

Pliny's discussion about Asclepiades' insistence on deduction from first principles (his *stocheia* and *dogmata*) points to another Greek writer, Theophrastus of Eresus. However, where Asclepiades employed questionable methodology and his background in rhetoric to exploit the people of Rome, Theophrastus was philosophically trained and relied on observation for his inferences. In his two extant works on plants, *Historia Plantarum* (*HP*) and *De Causis Plantarum* (*CP*), Theophrastus mimicked Aristotle's technique of classifying animals, as seen in *History of Animals*. He also employed Aristotle's technique of delineating the shared and common characteristics of animals in a secondary work, by identifying and classifying plants in *HP*, then describing their commonalities and patterns in *CP*. Part of the issue that both Galen and Pliny had with Asclepiades was that he either skipped the first step and went straight to the second, by trying to provide the underlying patterns of diseases and their plant remedies, or he reversed the process and tried to move from common characteristics and first principles to the individual classification process.[67] In contrast, Theophrastus began the *HP* by advocating for the study of each individual feature of a plant, physical or otherwise:

> One must take into account the differences in plants and also their nature, as these concern their parts, their qualities, their origins, and their life cycles.[68]

Theophrastus made it clear that the proper method for laying out and classifying individual plants, or more specifically, the differences between the various natural states of plants, is to take into account their parts as opposed to the whole and their conditions, their *pathe* (what happens to them). However, this is not enough. One needs also to consider their origins[69] and their life span, their *bios* (mode of living). Theophrastus spent the remainder of Book 1 going through the ways in which plants differ in these classifications; how trees differ according to their leaves, seeds, bark, growth habits, etc., and so on for the other kinds of plants. In Book 2, he discussed the various ways in which plants are propagated and their various origins:

> The origins of trees and plants in general are either spontaneous or from seed, from a root, or from a cutting, from a branch or twig, from the trunk itself, or even from cutting the wood into small pieces (for some are also produced in this manner).[70]

This is a literal application of the theory of first causes, which both informs and constitutes Theophrastus' methodology. Here, individual plant descriptions serve as instances of praxis in support of a theory. Rather than separate out the plant descriptions from his explanations of the ways in which they differ according to parts, origins, growth habits, etc., Theophrastus incorporated individual plants as examples. For instance, while discussing the ways in which plants can suddenly change their form or growth habit, he explained that celery (*selinon*) is affected by the way in which it is cultivated. After the seeds of celery are sown, if they are stepped on and rolled into the earth, the plant will grow in curly.[71] In this way, Theophrastus managed both to describe the cultivation and general characteristics of celery and to verify the premises underlying his methodology. Plants can be produced from seed, but their growth habits can be manipulated after the fact. In the absence of Asclepiades' *Precepts*, one might therefore substitute them with what we know of Theophrastus, since their methods of approaching herbs have a similar methodology. However, Theophrastus was not a practicing medic, but a philosopher and the successor to Aristotle. His status as a Peripatetic places him in opposition to Asclepiades' original training in rhetoric. Theophrastus was engaged in philosophical inquiry, while Asclepiades was engaged in the art of persuasion. Nevertheless, Pliny's reactions to both Greek thinkers can be compared and contrasted.

Pliny referred to Theophrastus no less than seventy times at various places in the *Historia Naturalis*, especially in regard to stones, earth, natural wonders, and buildings, with a large proportion in Book 36, where Pliny cited him as an authority on minerals.[72] Some of these citations demonstrate both how Pliny interacted with Theophrastus as an ancient authority and how inconsistent nomenclature creates confusion, even between two languages. For instance, in Book 26.63, Pliny described how many people believe that the roots of the plant *Satyrion* have aphrodisiac qualities if consumed. When Pliny cited Theophrastus' claim that this plant can allow for copulation seventy times in a row, he also noted that Theophrastus did not explicitly name *Satyrion* and did not assign name and genus (*nomen genusque*) to it. In this case, Pliny did not appear to question the plant's known aphrodisiac properties, simply the fact that Theophrastus, a serious author, had clearly exaggerated these effects. While Theophrastus parroted the idea that a certain plant had copulative powers, he did not specify name or genus, which meant that the plant could not be identified. At *HP* IX.18, Theophrastus said of *thelufonon* (wolf's-bane) that it is called *scorpion* (Scorpion-plant), because its root looks like a scorpion.[73] He listed some of its supposed effects, such as killing an actual scorpion, which can then be revived with white hellebore. Theophrastus then said that myths (*mythode*) are not composed for no reason. In other words, there may not be philosophical truth in these fantastic stories, but there is some purpose, some logic underlying their existence. Where Pliny denigrated Asclepiades for his poor methods, lack of experience, and questionable motives, he begrudgingly admitted that there could be something of value in his *Precepts* precisely because Theophrastus, whom Pliny respected, sometimes indulged in similar tales. However, we cannot assume from this that Pliny was a proponent of such stories. He was still cautious about the physical effects of plants, explaining at 25.6 how he would avoid delineating the properties of poisonous herbs, or plants whose use or ingestion he considered "criminal" (*magica portenta*). In particular, Pliny assigned love potions and abortive plants to this genre, declaring that faith in such things must be condemned.[74]

This brings us back to the moral component of Pliny's critique of Asclepiades and others who withheld knowledge of plants. Pliny distinguished between healing plants and healing physicians. A physician must diagnose and heal, not experiment on patients by testing out hypotheses, using innocent plants to do so. The healing properties of plants simply are what they are. Even if there is debate among physicians and healers regarding the correct use of a plant, its status as "medicinal" stands and is external to the field of medicine. A plant's healing properties are a part of its nature, regardless of their exploitation by physicians. This is reflected in the variety of adjectives and other modifiers that Pliny used to describe a plant's healing properties. On the other hand, when Pliny discussed physicians, who were supposed to deploy plants as medicine, the vocabulary of healing is noticeably absent, except where he took the time to point out their failure to do their job. For example, at 26.4 Pliny told of Laecanius Bassus, who perished from carbuncles thanks to the ignorance (*inscientia*) of his doctors and at 29.5 cited a popular, sarcastic funerary inscription, to the effect that someone had died because of the crowd of physicians (*turba se medicorum perisse*). At the beginning of Book 29, Pliny declared that he needed to say something about the art of medicine itself and provided a brief history of the field, starting with the Greeks. Throughout, he criticized not only their methods and tenets, but also their manipulation of knowledge of plants and their willingness to bilk their clients, who did not know any better. Moreover, many physicians were more concerned with establishing themselves via the novelty of their remedies than by their efficacy.[75] Referring to Marcus Cato (the Elder) at 29.8, Pliny stated that the elders of Rome did not condemn medicine itself, but rather the profession of medicine, because physicians charged a fee for good health. Pliny claimed that the Romans were so disgusted by this system of paying for the privilege of living that they banned Greeks from the city, including physicians (*excepisse medicos*). Because of this negative experience with Greek physicians, the Romans did not pursue the art of medicine and the few who did study it were literally deserters to the Greek side. The result was that the Romans became susceptible to the lies and deceptions of medical quacks. Because they themselves no longer possessed the requisite knowledge of the art of medicine, Romans were easily manipulated and those who called themselves "physicians" had *carte blanche* to test out their theories on them. "They learn from our dangers and conduct their experiments with our deaths, and to the physician alone there is absolute impunity to kill a man."

In Book 25, Pliny's discussion about the lost knowledge of the Greek physicians was centered on the custom of not sharing one's information. In Book 29, he focused on the immorality of the practitioners of the field, but Pliny was making a further, subtle distinction between experimentation and experience. The Greek physicians not only were exploitative and charged immense fees for their expertise, they were working from first principles and a system of deduction. Unfortunately, this meant that when they applied their theories, they did so from a philosophical-logical perspective, according to which their remedies *should* have worked, but without long years of experience in the field they were in effect experimenting on their patients. This is not the same as the knowledge of plants and medicine hard-won by those dwelling in the countryside. For Pliny, although the Greeks were to blame for taking advantage of others, the Romans deserved both the treatment they got and their current inability to heal themselves. The Romans put themselves into a dangerous situation by letting others do the hard work of learning about medicine, hence the need for a Roman to take up the task of not only learning all there is to know about medicine, but verifying remedies and writing everything down in Latin. At 19.62, Pliny claimed that "the true nature of a plant cannot be fully understood apart from its medicinal effect." For Pliny, there was a connection between regaining lost knowledge of the field of medicine and knowledge of plants. Likewise, if it was paramount to write about medicine in Latin for Latin speakers, it was also preferable to write about plants in Latin. Once the Romans had regained this knowledge, they would, however, need to construct their own mores and best practices. Pliny made it abundantly clear that the primary function of a physician is to heal, and secondarily to share their knowledge. If physicians had followed these two precepts, Pliny would have been able to use the same language of healing for them as he did for plants.

Conclusion

Three features of Pliny's text can be identified as components of a reference sphere for Otto Brunfels' 1530 herbal:

1. His use of binomial nomenclature, including his concern for accurate, descriptive names that assist with identification of plants.
2. His criticism of the customs of naming plants after the "discoverer" and of refusing to pass one's knowledge down to posterity.
3. His rejection of the Greek method of reasoning from causation or first principles, in favor of learning from experience.

Within the reception sphere, all three also constitute transformations, in and of themselves, of the ancient authorities whom Pliny cited. Pliny's moral stance on healing is not as easily located in the same reception sphere. He had declared more than once in the *Historia Naturalis* that he was the first to describe plants, especially their medicinal properties, in the Latin language. He began Book 29 with the following declaration:

> The nature of remedies, and the multitude of those that are present to hand and omitted, force me to say more about the art of healing itself, although I am not unaware that before now, no one has collated it in the Latin language, or that the judgment of all new things is uncertain.[76]

If indeed Pliny was the first to write about certain plants in Latin, his text can be seen as the *initium* and genesis of botanical Latin. Again, at 12.31, he said of the Frankincense tree that its appearance had, as far as he knew, not yet been described in Latin. The strict truth of these claims is not the point.[77] What matters is that they support the idea that it is possible to trace the ways in which the Latin language was used to describe and name plants and thereby provide a starting point for analysis of the slow progression to the strict rules of modern botanical Latin.

2

Otto Brunfels' Reception of Pliny

Botany and botanical Latin share a number of features with the botanical books of Pliny the Elder's *Historia Naturalis*, in which he critiqued the methods and traditions of Greek botanical writers, argued for knowledge (*cognitio*) gained from experience, and insisted on descriptive nomenclature. These characteristics share a focus on the relationship between knowability and nomenclature, and between language and knowledge. Pliny made much of the connection between plant nomenclature and discovery, implying that a plant name that does not provide information for identification purposes is "unknown" and "bare." Knowledge of plants is dependent on a plant's discoverability and on the successful transmission of information about it. Knowledge is therefore a product of good nomenclature, because poorly named plants cannot be correctly identified. The *Historia Naturalis* functions as a reception sphere, interacting with and transforming earlier botanical authorities, especially Theophrastus and the Greek tradition more broadly. In turn, they serve as a reference sphere for the German physician Otto Brunfels' 1530 text, *Herbarum Vivae Eicones* (Living Images of Herbs) for which the ancient natural historian serves as the primary ancient authoritative inspiration. Pliny's stance on naming and knowing plants therefore provides the background for the development of botany and botanical Latin in the sixteenth century. Brunfels appropriated Pliny's ideas and transformed them into an idiosyncratic herbal that was pivotal for the development of botany. The ideals of intelligibility, knowability, and discoverability, as Pliny framed them, find new expression in Brunfels' herbal, which in turn influenced another sixteenth-century physician, Leonhard Fuchs (1501–66), author of the 1542 *De Historia Stirpium* (An Inquiry into Plants). From Fuchs, the continuum of botany and botanical Latin is much more straightforward, with herbalists, botanists, naturalists, and other practitioners citing and responding to one another in the lead up to major points on the continuum, such as the classification work of Swedish botanist Carl Linnaeus in the eighteenth century and the extreme formalization of botanical Latin in the twentieth century. Both Pliny and Brunfels were instrumental to the development of the field, through their contributions to nomenclature, visual communication, textual stabilization, and their views on intelligibility, knowability, and discoverability.

Plinian Reception: The Reference Sphere

Otto Brunfels (*c.* 1488–1534) was a German physician at Bern who published between 1530 and 1536 three volumes of personal observations about German plants,[1] in which he provided both their Latin nomenclature and their names in the vernacular. Brunfels' original plan for the book had been simply to map the plants identified by the ancients onto plants known in Germany, to serve as a reference-point for other physicians and herbalists.[2] Brunfels' publisher, Johannes Schott, employed the artist Hans Weiditz (1495–*c.* 1537) to produce realistic illustrations of the plants. For Sprague (1928), this is precisely the feature that rendered Brunfels' herbal crucial to the development of botany as a science,[3] because Weiditz's woodcuts made identification of plants a far more precise process than it had ever been.[4] The botanical illustrations that accompany the text are often treated as the sum total of Brunfels' contribution to botany and while he is frequently discussed in relation to the history of art, in the history of science he is generally considered unoriginal and insignificant. The twentieth-century botanist-historian Agnes Arber (1912) said of Brunfels that his import in the history of botany lay in Weiditz's illustrations, "which are superbly drawn though with no sacrifice of realism to artistry."[5] Yet Brunfels' herbal functions as a reference sphere in which the descriptions of plants transform the classical Latin of Pliny and his rough contemporaries, and the amalgam of medieval Latin and the vernacular that characterizes the herbal tradition. The Latin descriptions of plants in Brunfels actively transform these forms of Latin and contribute to the development of formal, modern botanical Latin.[6] The methodology of *allelopoiesis* underlies the argument that the Latin of Brunfels' herbal shows the markers of classical and medieval "botanical" Latin, as well as the formal botanical Latin that began to develop in earnest with Leonhard Fuchs' 1542 *De Historia Stirpium*.

In the late fifteenth century, following the publication of the *editio princeps* of Pliny's *Historia Naturalis* in 1469 and the 1470 Roman second edition, there was considerable consternation among the Italian humanists about the quality of the printed text.[7] Initially, their concerns were focused on the preservation and accuracy of Pliny's text, but this soon transformed into debates about Pliny's own authority as a natural historian. In 1492, Niccolò Leoniceno (1428–1524) published a tract about the errors in Pliny in response to the classical scholar and poet Angelo (Agnolo) Ambrogini (Politian, 1454–94), with whom he had been engaged in an epistolary debate regarding Pliny and the errors contained within his text.[8] Leoniceno believed that Pliny mistook the Greek word *kisthos* (rock-rose) for *kissos* (ivy)[9] and had at some point written about this to Politian.[10] In a 1491 response letter to Leoniceno, Politian defended Pliny. He did not entirely agree with Leoniceno's analysis of Pliny's treatment of the Greek names, but in his letter, he granted that such mistakes, if made, could "render a doctor more dangerous than a disease," highlighting the importance of correct identification and nomenclature. In 1493, Ermolao Barbaro (1454–93), a Venetian humanist and ambassador to Rome, produced a work called *Castigationes Plinianae*, in which he claimed to have "healed" some 5,000 *vulnera* (wounds) in Pliny's *Historia Naturalis*.[11] In addition, Barbaro also authored a *Corollary* on Dioscorides.[12] Barbaro soon became the foremost authority on Dioscorides and Pliny and his emendations of the latter were incorporated into the 1497 Venetian edition of the *Historia Naturalis*.[13] The printing, editing, and correcting of Pliny's text was seen as literally pathological by the humanists, who borrowed medical terminology to draw attention to the importance of establishing a stable, authoritative version of Pliny's text.

Brunfels was heavily influenced by this debate and cited these same humanist scholars as his own authoritative sources, on botanical and Plinian matters both, especially Barbaro. Since they were neither botanists nor physicians, their influence on Otto Brunfels' work must derive from their connection to Pliny's text and the work they did to correct it. Brunfels' main project involved identifying German herbs with medicinal benefits, and secondarily to map the German plant names onto ancient Greek and Latin plant names, a process of verification. However, Brunfels did not blindly follow the authorities, especially if they were not ancient, and frequently inserted his own opinion into debates about a herb's nomenclature and characteristics. Given this concern for accuracy, the work of philologists such as Ermolao Barbaro struck Brunfels as pertinent to his own project. Moreover, in light of the connections made by Pliny between knowability and nomenclature, exactness in the *Historia Naturalis* takes on a new import: if there are mistakes involving nomenclature, our knowledge of the plants in question becomes unstable. The issue of knowability expanded in the era directly preceding Barbaro's *Castigationes*, in the thirty odd years following the publication of the first printed edition of Pliny in 1469. In this time frame, the idea of a plant's knowability extended beyond the confines of nomenclature to the stabilization of the "authentic" reference text. By stabilizing Pliny's text, the facts contained within it were also codified, which led to the transmission of more reliable "knowledge" about plants. This is of course somewhat ironic, since Pliny had argued so emphatically in favor of experience over book learning. For him, the knowability of a plant would never have been contingent on the stabilization of a particular written text, because the extent to which a plant is known is determined by the extent to which an individual has experience with the plant.

In addition to placing far greater emphasis on authenticating text and author, the humanist scholars themselves began using the language of healing, as though the *Historia Naturalis* were a physical body in need of medical intervention. This metaphor appealed to Otto Brunfels, himself a practicing physician, and he used it in his own work, although he applied it more broadly than to one ancient text. While Brunfels' appropriation of the healing trope cannot be said to have caused the development of the more systematic use of the language of botany, it was a crucial component of a broader discourse that resulted, eventually, in formal botanical Latin. Botanical Latin rejected the language of healing, in part because of its alignment with these humanist projects. Likewise, the results of the humanists' work, which included stabilizing plant names, was of professional interest to Brunfels, who, as a physician, appreciated the practical benefits of such projects. The reactions of humanist scholars to the two earliest printed editions of Pliny form their own domain of discourse and reception sphere and interact with and transform the features of Pliny previously discussed. Quite simply, it is precisely because Brunfels cited the emenders and defenders of Pliny that we can treat them and their own reception of Pliny as an intermediate locus of transformation; they were influential not only on Brunfels but also on the development of botany as a science. In addition, Brunfels' Latinity is liminal. It is not just a stepping stone on the way to twentieth-century formal botanical Latin, it is transitional and hesitant, but also highly influential and transformative, the product of his desire to identify and describe all of the known medicinal herbs in Germany and his frustration at not being able to borrow the language and vocabulary of external authorities in order to do so.

Alii viderint (Let Others See to It)

As a case study based on the methodology of *allelopoiesis*, Brunfels' herbal is embedded in a reception sphere in which it interacts with Pliny's text, then reconstructs and transforms it into something new. Brunfels' remarks about his authoritative sources function as data points for fleshing out the reciprocal relationship between his herbal and its sources. It is therefore crucial to examine the manner in which Brunfels both discussed Pliny *et alios* as authorities, and placed himself in the role of an anxious disciple. Brunfels employed a rhetoric of uneasy reluctance throughout the herbal, which, while a trope, is also indicative of his feelings of inferiority to the authorities on whom he relies. He expressed frustration with the *Pandectae*[14] and with Simon Ianuens, two earlier medieval medical authorities,[15] while quoting liberally from Ermolao Barbaro, Niccolò Leoniceno, and Pliny, among others. Brunfels' issue was primarily with those who had contributed to the corruption of Pliny's text, whether in the herbal tradition or via poorly edited printed editions. Despite these issues with the transmission of Pliny's text, Brunfels took from it certain markers of classical botanical Latinity, such as genuine concern over issues of nomenclature, the promotion of what were known in the herbal tradition as "simples" over compounds, and a focus on empirical observation. Brunfels transformed these elements of the reference sphere in his original botanical Latin entries.

In the course of their work, Brunfels and the artist Hans Weiditz encountered an unforeseen problem. The original plan had been to simply track down medicinal plants known in, endemic to, or propagated in Germany and to catalog the descriptions of these plants as found in the authorities. However, as their work proceeded, Weiditz, who was out in the field, encountered a number of plants that Brunfels could not find in the authors on whom he relied. Because of this, Brunfels' method switched to include and compile plants in the order in which he received the specimens and illustrations from Weiditz. He began the herbal with plants for which he could provide the known Latin, Greek, and German names, along with passages taken from Pliny the Elder, Dioscorides, and Galen. Although he considered Pliny the "father of antiquity" and exhorted everyone to take up his auspices,[16] Brunfels also frequently used Barbaro as a reference and claimed that he owed his own extensive knowledge (*cognitio*) to Barbaro's work on Pliny and Dioscorides.[17] Whenever it was necessary to consult other sources, Brunfels looked to Niccolò Leoniceno, the "greatest and foremost doctor in this day and age," and to the counselor of law Pandolfo Collenuccio (1444–1504), whom he described as so learned in plant lore that the whole crowd of medics that came after him came to metaphorical blows in their efforts to surpass him.

Brunfels provided two rationales for his work. First, he claimed that the herbal tradition, which had long provided concise assistance to medics, had by his own time collapsed. Second, in the absence of readily accessible, reliable herbals, Brunfels saw it as his public duty to fill the void and to bring to the light of day and to public knowledge crucial information that had been lost.[18] This appeal to the common good is not unexpected: herbs have well-known medicinal properties. What is remarkable about this rationale is that Brunfels identified a need based on the collapse of what he called *haerbariae* (herbalists) and, with false modesty, indicated his intention to fill it by means of his own "inferior hands." He hoped to take up where the herbalists left off, but resolutely refused to declare that he had achieved this. The Latin pronoun used to refer to what he hoped to call back into the light (*eam*) is frustratingly imprecise, with no clear antecedent. It could refer to a missing term such as *traditio* (tradition), to accompany the "collapsed" herbalists, in which case the medieval herbal tradition would be what Brunfels intended to resurrect. However, *eam* could also refer to the herbal itself, along with its subject, the humble *herba* (herb). This option supports the argument that for the early botanists, the medieval herbals had been a source of anxiety because of their errors, their incompleteness, their lack of an appeal to authority, and their anonymity. This anxiety has a parallel in the debates about Pliny's text: if the manuscripts and printed editions are error-ridden and incomplete, then the information in them becomes questionable and uncertain. Brunfels was both genuinely expressing and exploiting this anxiety, thanks to which he intended to restore the modest herb itself as an object worthy of inquiry. Either one of these options is compatible with the crucial point, that there is a parallel between Brunfels' proposed revival of the study of herbs and the devolution of the Latinity of herbals and of European literature more broadly. For Brunfels, the humble herb had lost its prestige, which was due in part to the carelessness of the herbalists themselves. Not only did they make critical errors, but their Latinity had become by the fifteenth century simplistic and of poor quality. For example, a thirteenth-century herbal says the following of the herb *Betonica*:

> Herba betonica viride bene constricta in mortario et supposita mirum sanabit.
>
> The herb *Betonica*, if ground up in a mortar while still green and given as a suppository, will heal marvelously.[19]

This is a typical herbal construction, involving a future indicative verb (*sanabit*) accompanied by attributive past participles (*constricta* and *supposita*) that modify the subject of the verb, *Betonica*. This general structure, in which the participles point to some method of gathering, treating, and preparing a herb, which then is supposed to result in the healing action of the verb, is so ubiquitous that it could be said to be incantatory in the same way that religious rites are. A later herbal from 1484, the *Herbarius Moguntinus*, is more sophisticated syntactically, but betrays the same formulaic tendency:

> Betonica est calida et sicca in [terciobus] virtutem abstergendi et incidendi grossos humores ideo valet ad frangendum lapidem renum et vesice et provocat menstrua. Et valet in doloribus lateribus.[20]

> *Betonica* is readily available, and when dried in [thirds] has the virtue of dispelling and eliminating excessive fluid; and for this reason, it is useful for breaking up kidney and bladder stones and it stimulates the menses. It also is beneficial for pain in the flanks.

Apart from the provision of the plant's German name and apart from an accompanying woodcut, which is quite basic, no identifying information is provided. Rather, the text immediately launches into a catalog of the herb's medicinal properties. The Latin has a simplistic, repetitive feel to it, despite the presence of gerunds, participles, and ablatives. However, despite Brunfels' complaints about the herbal tradition, this entry shows that the herbal lexicon had clearly already changed, with the inclusion of more technical vocabulary. This is a crucial development, as it shows that by the late fifteenth century, after the philological work of the Italian humanists, the importance of overall precision had made its mark even on the herbalists, who expressed their growing concern for exactness in a new and expanded lexicon.[21] Rizzo (1973) discusses several terms that involve Pliny and the Italian humanists, including those that emended or collated his works, such as *emendo* and *textus*. Rizzo avoids discussion or comparison of classical versus humanistic uses of medical-herbal terms like the verb *sano* (to heal), or of the terms *herba*, *herbarius*, *botanica*, *radix*, *flos*, etc. Rizzo's work purports to assist in understanding the technical terms used in the philological activities of the humanists,[22] who were serious about stabilizing Pliny's text, but not so concerned with the precision of the Latin of its contents. Nomenclature and botanical terminology were not of concern to them. Yet if a "study of words is also a study of things,"[23] then the words we use are an indication of the things about which we care.

The association of herbals and herbs with a loss of quality had real implications beyond philology. In 1542, Leonhard Fuchs would lament the fact that in his time, knowledge of herbs and their medicinal uses was confined to superstitious, uneducated women.[24] The long association of herbals with the non-scholarly community underlined the need for establishing a more "respectable" milieu for the study of plant material. Brunfels therefore took up the theme of restoring the study of herbs to its former glory, but was careful to position himself as a mere compiler and consumer of more qualified authorities. He declared that he wished to respect the Hippocratic oath to do no harm and to not lead others astray.[25]

However, he did not see himself as equal to the many learned medics whose expertise he needed in order to effectively critique the established authoritative texts. Even if he were able to engage in such a critique, he would not have the authority or duty to do so. As a humble physician, however, he did think that he was qualified to compile a book of herbs and in so doing, recall his peers' attention to the value of knowledge of plants.[26] The use and knowledge of herbs were therefore Brunfels' main goals in writing his herbal.[27] Yet botany as a science is unconcerned with the practical uses of plants. Brunfels' project is therefore liminal, with stated purposes that are both practical and theoretical. The evolutionary nature of the text is evident not just in Brunfels' approach to herbs but in his use of binomials and in the Latin that he uses to describe plants. It is, furthermore, evident in his prodigious citation of earlier authorities.[28] Where the herbalists had frequently gone rogue on nature's productions, either making potentially deadly mistakes or outright making things up, and where later botanists relied on their own observations of plant life, Brunfels in turn relied heavily on the consensus of his predecessors, as though he were sure enough of his own skills to identify the need for a more modern herbal revival and a renewed interest in the uses of herbs, but not confident enough to take on the full responsibility. Herbs can, after all, be deadly.

Yet herbs are also a necessary component of doing medicine. For Brunfels, Pliny became the object of near-religious fervor, in part because Pliny recommended simple herbal remedies over complex mixtures. The promulgation of these concoctions by the pharmacists, apothecaries, and misguided medics of his day constituted a fraud on everyday people, who could find "simples" for free in their gardens. Brunfels provided a lengthy quote from Book 24 of the *Historia Naturalis* to support his argument that herbs are *de facto* readily available, that there is no need for fancy concoctions from abroad. Pliny claimed that simple herbs were the only things that nature had seen fit to provide to all and sundry as remedies, being easily found, free, and part and parcel of daily life. In fact, the pharmacists' frauds were exposed when the apothecaries and their "deplorable workshops" came along, promising health at a price via complex concoctions instead of simple herbs. "At once, mixtures and mysterious compounds are reeled off."[29] Because of the complexity of these remedies, it became necessary to seek out ingredients from abroad, from Arabia and India, which Pliny saw as a travesty, since anyone could look in their kitchen garden for a herbal remedy that required no skill to prepare.[30] The skills of medics were no match for the effective simplicity of the common garden herb and Pliny's own sources being closer to home than Arabia and India, his remedies were more naturally pleasing.

Pliny and Brunfels were both reduced to fretting over the lost wisdom of the ancients and the deleterious effects of a more modern concept of "skill." The concoctions sought from abroad were overkill, since garden herbs would suffice, but more than that, they were dangerous because they were unfamiliar and their ingredients unknown, which is why Pliny referred to them as *compositiones inexplicabiles*. Relying on religious terminology, Pliny lamented that by seeking greatness, the Romans had lost the "rites" of the ancients and were now ruled by their own skills (*artes*). Brunfels' appropriation of Pliny's words to express these concerns therefore marks a first transformative process within his reception sphere. Brunfels appeared to enumerate the same frustrations as Pliny, but in fact there was a crucial difference in the substance of their complaint. Pliny's words constitute rhetorical asides easily overlooked in the sheer magnitude of the *Historia Naturalis*. His arguments for the value of knowledge about medicinal plants are important, but they are hardly the main focus of the work as a whole. Brunfels had the intervening 1,500 years of the herbal tradition with which to contend. Where Pliny lamented what had been lost, Brunfels agonized over what had been corrupted. Pliny's thoughts on these matters clamored in Brunfels' mind: distrust of medicines prepared with plants that are not found at home and are therefore not empirically verifiable; reverence toward the ancients; and frustration at being confronted with new plants at home.

Brunfels warned his readers about the dangers of plant knowledge being concentrated in the hands of the few individuals[31] and reiterated his desire to recall practical uses of plants and to reinstate an uncorrupted form of the ancient texts he revered.[32] Brunfels would have had nothing to say were it not for the work already done by the ancients. The only substantive contribution Brunfels could claim was the fact that he was bringing what the ancients said back to the attention of those in the sixteenth century. Despite his performance of humility, Brunfels was in fact engaged in an *allelopoietic*, transformative process involving not only Pliny, but all of the botanical authorities who came before him. By collating their writings on each of the herbs, as well as the opinions of Pliny's emenders, Brunfels tackled head-on the confused muddle resulting from the herbal tradition. He verified what is correct, settled name disputes, and ensured, most importantly, that the consumers of his herbal would not make any fatal or harmful errors. The medieval herbal tradition was one of the key catalysts for this transformation.[33] The only information that was "pure" enough to be passed down was found in the ancients, and medical practitioners would become better at their job by referring to these authorities, not by looking abroad. The medieval herbalists' writings were nothing more than robbery (*alienum furtum*) and loan words (*asciticia*). They had lost the splendor of the ancients and developed a way with words that was now insipid and befouled. As a result, Brunfels found it necessary to highlight the works of the true masters of the skill of plant medicine.[34] He hoped that the vice that he identified in so many authors, medical practitioners, and other purveyors of herbal medicine would be absent in his own work.

Brunfels' theories and methods can be seen in his entry for the herb *Personata*.[35] He cited *Historia Naturalis* 25.58, where Pliny explained that the plant known in Latin as *Personata* was called *Arction* by the Greeks. However, there was some confusion surrounding the nomenclature, since *Arction* is written in some printed editions and manuscripts as *Arcion*, or even as *Echion*, which Ermolao Barbaro discussed in his *Corollaria* on Dioscorides. Barbaro argued not only that *Arction* and *Echion* were different plants, but that *Arction* and *Arcion* were also taken as different in Dioscorides, whereas Galen saw no difference between them. Barbaro also posited that where Pliny used *Personata*, he in fact meant the plant called *Petasite*. To add to this tumult, *Arction* was also known in Latin as *Persoluta*, *Bardana*, or *Lappa maior*.[36] Brunfel's inclusion of this debate surrounding *Personata*, and whether or not Pliny mistook it for another plant, serves as a useful reference for the reader who is interested in the plant's nomenclature. But the issue is confusing and given the primarily medical purpose of the herbal, it is unclear why any of his readers would need to be privy to the complexity of the debate. Brunfels betrayed a more sophisticated motive than the publication of a medical herbal by cross-referencing with one another the fifteenth-century Italian humanist criticisms of Pliny's take on *Personata*, and what name he himself used, as opposed to later scribal emendations. Because Pliny was blamed by the humanists for reporting what others thought, in Brunfels' mind, any responsible discussion of the plant *Personata* must include these debates. At the same time, by laying all of this out, he himself sidestepped the same accusation of irresponsible reporting leveled at Pliny. In addition, it is plausible that a reader of Brunfels' herbal would also have had access to Pliny, Galen, Dioscorides, and Theophrastus. If, as Brunfels hoped, his herbal became widely popular, his readers could avoid the danger of mistaking one herb (*Echion*) with another (*Arction*, or *Personata*, or *Lappa maior*), and therefore potentially harming their patients. Ultimately, Brunfels approved of Barbaro's take on the situation, and Collenuccio's, and reestablished Pliny as trustworthy.[37] When Pliny said *Echion*, he meant *Echion*, not *Arction*, whatever errant manuscripts and editions might say. Brunfels' insertion of his own view, not only on the nomenclature and identifying features of the herb *Personata*, but on the fierce debate surrounding it, is singular even within his herbal. There are other cases of herbs regarding which Barbaro and Leoniceno took issue with Pliny and for which Collenuccio took up the defense, such as the herb *Betonica*,[38] but Brunfels refrained from inserting himself into the debate. In this case, the sheer confusion of the nomenclature and the question of the number of types of *Personata*, combined with the possibility of misidentification and misuse in medicinal concoctions as described by well-known medical authorities, were all sufficient to goad Brunfels into the expression of an actual opinion.

Brunfels' Latin: Transforming the Reference Sphere

Long overlooked in favor of Weiditz's illustrations, the Latinity of Brunfels' herbal is just as valuable for understanding the early development of botanical Latin as its woodcuts are for the development of scientific illustrations. The very feature of Brunfels that has drawn the most criticism from scholars of botanical history, namely, the unoriginality of his reliance on prior authors, is in fact one of its most important components. Brunfels aggregated his sources and in so doing, painted a picture of the intellectual state of the study of plants in Western Europe. Brunfels tried to reconstruct the authoritative picture of the herb *Personata* and in the process, appropriated it into his reception sphere. This appropriation comes further into focus when we consider that in many other cases, beyond the herb *Personata*, Brunfels inserted what he calls a *Iudicium Nostrum* (My Opinion) in which he attempted a conclusion regarding particular debates about a particular herb. These *iudicia* are significant because they are Brunfels' reluctant transformations of the markers of Plinian botanical Latin into something more modern and more precise. While Brunfels implied that he would have been happy to simply cite his sources, circumstances were such that he was forced to resort to new ways of describing plants. There are also several herbs throughout the text that Brunfels could not find in his sources and therefore described on his own. These are German plants that did not have identifiable counterparts in classical antiquity and had to be described in original terms and vocabulary. Appropriating Pliny's term *nomina nuda* (bare names), he referred to them as *herbae nudae* (bare herbs). His reluctance to describe these "bare herbs" goes beyond the humility trope. Even if he were able to map the German nomenclature onto Latin names found in the medieval herbals, Brunfels would not have found their entries worthy of copying. It was important to him that whatever information he included in his own herbal was comparable to the ancients and worthy of being set alongside them.

Čermáková and Černá (2018) have focused their research on the *herbae nudae*.[39] In particular, they distinguish Brunfels' descriptions of these *herbae nudae* from later descriptions of plants "discovered" in the New World, relying on Ogilvie's (2006) classification of Renaissance botanists into distinct generations.[40] Into the first, Ogilvie places Niccolò Leoniceno and Euricius Cordus,[41] who were not so much concerned with scientific verification as with a straightforward mapping of European plants onto those in Pliny and Dioscorides. The second generation involved phytography, in which real plants that could be found in nature and illustrated were mapped onto plants described in the ancients. The primary focus of this second generation was to verify the information found in the ancients and their projects were based on the assumption that the ancients did not have the information needed to engage in scientific verification of their subject matter. This resulted in the development in early botany of a "science of describing" the natural world with extreme clarity and precision, a crucial component of which was the use of realistic illustrations. Brunfels belongs to this second generation and his *iudiciae* and *herbae nudae* are therefore paramount in any analysis of the Latinity of his herbal. Čermáková and Černá argue that Brunfels' *herbae nudae* are devoid of the "rhetoric of wonder" that characterized descriptions by Europeans of "unknown" plants in the New World, because they were well-known to him. Rather than amazement, he expressed frustration at having to describe them. In doing so, Brunfels, the reluctant botanist, appropriated the Latinity of the authorities and simultaneously began to transform it into an original proto-botanical Latin.

Herbae nudae (Bare Herbs)

Whereas Brunfels wished to employ illustrations of unfamiliar plants in his herbal, he was not concerned with the actual description of them.[42] Brunfels' thoughts on these unknown plants, which he believed were either uniquely German or at least unknown to the ancients, can be found interspersed throughout the herbal. In keeping with his process of describing herbs in the order in which their illustrations arrived at his desk, he used the heading *Otho Brunnfelsius* for the first *herba nuda*, alternately called *Kuchenschell*, *Ruchschell*, or *Sacketfraut*, as if to highlight the fact that the authorities on whom he had relied for the prior section could have no say here. Rather than begin with his usual list of the known Latin, Greek, and German names, he editorialized, explaining that when he began work on his herbal, the plan had been simply to include an appendix of any plants that he could not positively identify in his sources. He had meant to provide full entries only on plants for which he could map the German name to the established Greek and Latin nomenclature, which are identifiable and used medicinally in Germany. But the continued appearance of plants that did not fit these criteria forced him to take a closer look at them, to "take the opportunity they provide."[43] The fact that Brunfels had encountered a number of German plants that he could not map onto the ancients caused him anxiety, because he could not refer to others as authorities.[44] In fact, there were other herbs prior to this one, which Brunfels had also been unable to definitively identify, but it was not until the plant *Kuchenschell* that he took the time to highlight the problem, betraying his lack of interest in pre-planned organization. The plants in Brunfels' herbal are literally presented to the reader in the order in which the illustrator captured and presented them to him. However, after so many unidentifiable specimens, Brunfels suddenly lost his patience.

This loss of composure is best seen by comparison with an earlier plant for which Brunfels listed the Teutonic nomenclature as either *Gulden Guntzel*, *Brimel*, or *Grosse Guntzel*. Likewise, the Latin nomenclature is so confused that Brunfels could not map the common names with any precision.[45] Dioscorides called it *Solidago*,[46] but others called it *Consolida media* or *Consolida minor*. In this example, Brunfels admitted that he did not know for certain the proper name for the plant. There do appear to be Latin proto-binomial names for the three known species of this plant, but because Brunfels was unable to map them onto the ancients, he instead mapped the varieties of flower onto the German names. Given the lack of authoritative data on the plant, Brunfels decided to simply list its medicinal properties: it heals wounds, dissolves blood clots, and remedies mouth ulcers. Here, Brunfels relied on his own experience as a medical doctor, using the verb *vidi* (I have seen) to explain *Consolida*'s known and alleged uses. In keeping with his self-deprecatory, humble trope, he was only comfortable saying this much about this particular herb because it fell within the realm of medicine. As a non-botanist, Brunfels had more to say about the plant's healing properties than its identifying characteristics. Yet, when presented later on with *Kuchenschell*, Brunfels suddenly had much less to say about its medicinal uses, although he was verbose about the difficulties of identifying it. Brunfels' concern was with healing and the unknown was of no interest to him, which explains his frequent use of the term *cognitio* to refer to his knowledge of the subject, a term borrowed from Pliny. Yet, despite his intention of relying on the authorities, Brunfels' project took on a life of its own. In the process of gathering illustrations of German plants, he found himself confronted with a stockpile of plants that could be of medical benefit, but which he could not verify by reference to any authority. The uncertainty involved in identifying these plants by name and verifying their efficacy was the impetus for his decision to make the attempt to do so. If there was any possibility that these *herbae nudae* could be used medicinally, he had a duty as a physician to try to do so.

When confronted with the herb *Kuchenschell*, he therefore explicitly stated that he had decided to delineate those *herbae nudae* with German names. He was hesitant to take up this task, but ultimately decided to include the *herbae nudae* because the artists and engravers (*delineatores et sculptores*) of the illustrations of more common herbs had been delayed. In order to keep the process of writing and printing the book moving, Brunfels was forced to turn to other herbs that were readily available for description. However, he was unsuccessful at procuring their Latin names in the inventories of either the physicians or the herbalists.[47] Brunfels' inclusion of these herbs in his own work could provide basic information about them to those who were more educated about such matters, such as their habitat and uses. His decision to include the *herbae nudae* therefore amounts to passing the responsibility of accurate description and identification over to others. Individuals who were more skilled in botanical matters might use Brunfels' entries to do further research and flesh out his contributions, which consisted primarily of their medicinal uses. Brunfels chose his terms carefully, always stressing his lack of skill in the process of exploring the works of the authorities. He lamented how so much was missing from his sources that he could not even identify the herb *Kuchenschell* in Dioscorides or any other ancients. Rather, his concern was with providing an opportunity to those who were more knowledgeable than he was about such things. He thought that these others would complete his catalog, which he assumed was going to be well-known. While he could not fully identify and describe the *herbae nudae*, he could say, from his own empirical observations, that they might be curative (*vulneraria*).[48]

Brunfels' citation of empirical observation as the preferred method of deriving knowledge of a plant's medicinal properties does not wholly correspond to his satisfaction with Hans Weiditz's illustration of the plant. He did appear to have examined a specimen, in order to offer a physical description of it. The truth is that Brunfels was not concerned with plants for which there were no known medicinal uses. For *Gulden Guntzel*, *Brimel*, and *Grosse Guntzel*, Brunfels was at least able to discuss the possibility that they are three species of the same genus, *Solidago* or *Consolida*. However, for *Kuchenschell*, he had no point of reference, which caused him such consternation that he indulged in a fit of pique and provided only a short paragraph on its medicinal uses. In this case, because he could not rely on the authorities and because he had decided to include the unknown plants whose illustrations Weiditz sent along to him, Brunfels was forced to provide all of the known information about the plant. But he stayed in his comfort zone and simply listed the medicinal properties:

> Secantur folia eius non sine acore quodam, ita ut oculos quemadmodum allium, aut cepe feriant. Illaque per alembicum destillata, aquam vulneribus mundificandis, curandisque utilissimam praestant, quae et putridam carnem erodant.[49]

> When the leaves of this plant, which have a certain bitterness, are removed, they sting the eyes, like *allium* (onion) does, or *caepe* (also onion). And when the leaves have been distilled in an alembic, they produce a liquid that is very useful for the cleansing and healing of wounds, and they also erode rotting flesh.

Brunfels was using a technique found even in Pliny; he compared the bitterness of the leaves to onions, which are notoriously noxious. Beyond this, however, his information is not very specific. We learn only that a decoction of the leaves can be used to treat wounds and clean out gangrene. However, *Kuchenschell* is only the first of two herbs filed under *Otho Brunnfelsius*. The second is called *Gauchblům* (Cuckoo flower) and while an illustration is provided, Brunfels could only be bothered to include two sentences about it. He stated that it was unknown (*incognita*) to the ancients, that it grows throughout Germany, and that some people believe that the pedicel, if washed in lixivium, has medicinal uses for foot issues.[50] Of note in each of these entries is the fact that Brunfels provided no information on how to identify either *Kuchenschell* or *Gauchblům*, apart from their illustrations. Their medicinal properties are derived from their leaves and certain specific parts of the stem or fruit. Thus far, Brunfels kept to the herbal tradition. However, he made no attempt to provide Latin botanical names for the plants, nor did he employ the binomial format. His Latin is devoid of anything that resembles character. It is clear and succinct, but displays no stylistic or identifying features, as though to highlight Brunfels' refusal to place himself in the number of learned men. Despite this, his Latin is more complex than that of the late medieval herbals. The sentences are straightforward, but they are just intricate enough, with just enough subordination, to distinguish the technique and style from the herbal tradition.

This Latin, nondescript as it may seem, is of great importance in the development of botanical Latin. In this context, it does not matter whether or not Brunfels was right in assuming that the plants were truly unknown. The imprecision of naming conventions since antiquity and the lack of illustrations accompanying ancient texts means that some plants considered by Brunfels and Fuchs *et alii* to be unknown to the ancients, were in fact known under different names and descriptions. For some, this is a larger issue, because the Latin used by these early botanists to describe *herbae nudae* can also be distinguished from the Latin used to describe another group of plants unknown to the ancients: those of the New World.[51] Moreover, in the couple of decades following Brunfels' work, other naturalists began to focus more on experience to flesh out their botanical works, which meant traveling more in order to examine plants in person. This turn toward experience can be attributed to the development of botany as a science and the corresponding turn away from ancient authority to scientific authority,[52] but it was not the impetus for Brunfels' description of the *herbae nudae*. Rather, he decided to include them, briefly, so that others could follow up on his descriptions and become authorities on them.[53]

Twelve years later in 1542, Fuchs identified *Gauchblům* as *Nasturtium agreste*, also known as *Hiberis*. Despite not wanting to position himself as an authority, Brunfels clearly had *cognitio* about herbs that Fuchs and others restated and rehashed. Brunfels also had opinions about the general state of knowledge about plants. He acknowledged his debt to the ancients, but lamented that their knowledge of plants had over time become garbled and incomplete. In particular, he criticized those who later on "corrupted and confused everything in remarkable and prodigious manner."[54] Specifically, Brunfels blamed the author of the *Pandectae* and Simon Ianuens, who really mixed everything up (*omnia miscuerunt*), because they exerted themselves more over producing huge volumes of work than in leaving anything useful or knowledgeable to posterity or investigating real knowledge of such things.[55] It was also due to the poor quality of these texts that Brunfels and his contemporaries were left with a tangle of confused, obsolete, and useless names. Brunfels was deprived of *cognitio* about ancient things, which these authors had either willfully misrepresented or at the very least neglected. It is in these critiques of the medieval herbalists that the rationale behind Brunfels' own herbal becomes evident: he intended for it to be a reliable compendium of accurate illustrations and sufficient, consensual information from a variety of authorities. Moreover, his reliance on the emenders of the printed editions of Pliny is also explained: Ermolao Barbaro and Niccolò Leoniceno did more than just nitpick the editing and printing of Pliny's text in the late fifteenth century. They also helped to reverse the corruption of Pliny's text in the herbal tradition throughout the centuries.

In the cases of both *Gulden Guntzel* (*Solidago* or *Consolida*) and *Gauchblům*, while Brunfels was unsure of their corresponding names in Greek and Latin, he was able to indulge in some conjecture. It took him a while to realize, as he finally did with the *Kuchenschell* entry on page 217, that a trend was emerging that he had to address. Prior to this, the first *herba nuda* was found on page 182, although he did not yet recognize the significance of his discussion about its nomenclature. This herb differs from the others in that Brunfels listed two species, *Perfoliata mascula* (*Durchwachß Månnlin*) and *Perfoliata foemina* (*Durchwaß* [sic] *Weiblin*). At this point, as he indicated, he was midway through the book, with various herbs being brought to him in no particular order.[56] Suddenly, a herb known as *Perfoliata* in the apothecaries (*officinae*) was presented to him. This herb caused Brunfels considerable anxiety because he could find out nothing beyond its Latin and German name[57] and he decided that the *Perfoliata* in question was a counterfeit, because the individuals who brought it to him were not, in his opinion, real authorities. However, one *pharmacopola* (pharmacist) named Lukas had shown Brunfels another herb with the same name, which was different enough from the other to remark upon.[58] This other *Perfoliata* might have been of the genus *Catapodium* or *Esula*. Moreover, its features fit so well with the plant bearing the German name that he was inclined to think that the herbalists who showed him the first plant were wrong.[59] The true German *Perfoliata* was the one described in Dioscorides and was kept by the medic Nicolaus Capitonus for healing broken bones and other wounds.[60] Brunfels was not able to discover any other herbs (*plura non comperi*) called *Perfoliata*.[61]

There are several remarkable features of this entry. In the first place, Brunfels was frustrated by the confusion surrounding the name *Perfoliata*, foreshadowing the nomenclatural confusion that caused later botanists such as Linnaeus to formalize botanical names. He also used the ablative case in his description of this second "Perfoliata," in such phrases as *coliculis scilicet foliola penetrantibus* (with little bumps penetrating the leaflets) and *altitudine dodrantali* (¾ foot in height). This technique later became standardized in modern botanical Latin where the diagnosis, which provides the distinguishing characteristics of the plant, is composed using either a string of ablatives hanging off of the species name or an indicative statement using the nominative.[62] In addition, Brunfels' entry on *Perfoliata* made novel use of scientific terminology. Considerable work has been done to map the terms used by the early botanists to modern terminology, such as Helen Choates' 1917 translation of the terms in Fuchs' glossary.[63] There is no corresponding translation for Brunfels, but considering their temporal and spatial proximity, he likely used the terms in a fashion similar to Fuchs. For instance, the term *herba* was defined by Fuchs as "a plant whose leaves begin from the root, grows without a stem, and often bears seeds on a stalk."[64] However, few of the botanical terms used by Brunfels in the above entry, such as *foliola* (leaflet), are in Fuchs' glossary. This makes Brunfels' terminology singular, informed by his status as a physician first and foremost. Fuchs was both an academic and a physician and this no doubt plays a part in the words they used to describe plants. Some words do appear in Fuchs' own entry on *Perfoliata*, however, such as *coliculus* (a small swelling like a flower bud). Fuchs specified how there are "small heads" at the top of the swellings, with "saffron yellow flowers and seeds that bear a form close to that of *Tithymallus*, which in the common tongue is called *Esula*."[65] This style, in which the sentence lacks a finite main verb but is accompanied by a matter-of-fact list of physical attributes, became over time more and more common in botanical descriptions. Fuchs' use of the term *coliculus* is surprising only in that it is not included in his own glossary at the beginning of the text, an indication that in the context of discussing a herb, a *coliculus* was so common and ubiquitous as not to merit comment. In the case of *Perfoliata*, therefore, Brunfels gave up on trying to precisely identify it, except to question whether the herb that is called *Perfoliata* by the herbalists was in fact the true *Perfoliata*.

Having dismissed *Perfoliata*, however, Brunfels was almost immediately confronted with another problematic herb, *Merzenblümen*. In this case, he attempted identification by reference to the parts of the flower, which were so similar to another species of *Narcissus* that Brunfels could only assume that the German name *Merzenblümen* referred to another species of it because he could not think of another genus that fit.[66] However, he refused to commit to this suggestion, hiding behind meek verbs such as *suspicio* (I suspect) and adverbs such as *fortasse* (perhaps). He was perfectly fine with being educated and corrected by someone better informed and was more interested in provoking those who were more erudite than in teaching others. Perhaps relying on the accompanying illustration to do the work for him, Brunfels concluded lamely that because the two species of *Narcissus* in question were so similar, the specimen in hand could be either one, although he decided to go with *Narcissus martius*. Considering how obvious and unhelpful this conclusion is, the point of the entry seems to be an *apologia pro se*. Brunfels admitted that he was willing to be proven wrong, should more information be found concerning this plant. If he were not busy with the work given him, to provoke those who are more learned, rather than to teach others, he would have overlooked his suspicion about *Merzenblümen*.[67] In other words, there are three choices for the conscientious herbalist with regard to plants with which they have little or no personal experience: they can pretend they do not exist; they can include them in their catalog and simply regurgitate lore regarding them; or they can make some attempt to identify them and reason through the maze of information found in their lore. Brunfels chose the last, in direct opposition to the medieval herbalists. Unlike Fuchs, however,[68] Brunfels was more grateful for the knowledge he gained about herbs from old women and herbalists, although he refrained from providing in-depth analysis of the *herbae nudae*. The sheer number of plant species described in other sources is a bounty, but it also introduced two issues for Brunfels. First, there was the problem of naming them all and second was the fact that he could not go into particularly deep detail about them. This is one of the primary reasons why there is such a dramatic difference between Brunfels' entries on plants identified in the ancients and his descriptions of *herbae nudae*. By citing ancient and medieval authorities, Brunfels was relieved of much of the labor of providing precise, accurate data. Where he could not rely on others, he balked at the work and insisted that others were better equipped to name, identify, and accurately describe the plants in question. This self-deprecation was apparently taken at face value by those who followed him and as a result, Brunfels' legacy is most often relegated to the quality of the illustrations done for him by Hans Weiditz.

Iudicia nostra (My Opinions)

There are several herbs for which Brunfels was compelled to provide his own opinion, always placed after the summaries of the authorities and titled *Iudicium Nostrum*. These entries differ from the *herbae nudae* in that, while some are unattested in the Roman authorities, they are found in the Greek writers, in particular, Dioscorides and Theophrastus. The issue with these herbs is therefore not one of mapping data sets, but making sense of the nomenclature, both current and obsolete, as found in the herbal tradition. This is why, faced with a plethora of names and species for the herb *Serpentaria*, Brunfels claimed that he "broke" under the strain of describing and depicting it (*depingenda et delinianda*).[69] He made this declaration well into a lengthy diatribe about the corrupting effects of Simon Ianuens and Pandectarius. Brunfels begged pardon from his readers, because he could not precisely portray the herb and all its names and features. He had managed to identify two species of the plant often found in gardens and cultivated places, with the common names *Gut Heinrich* and *Schwerbel*. The quality and accuracy of what Brunfels learned about them, however, remained for others to see (*alii viderint*).[70] These two words sum up

Brunfels' plan to leave useful information behind for other medical types and herbalists to name, identify, and verify. He used the subjunctive mood (*viderint*) and pointed to *vetulae* (old women) as his source, so that anyone who took issue with his description could direct their ire elsewhere. This, he thought, was the main difference between himself and Pandectarius or Simon Ianuens, who presented data about herbs with such authority that they lead others into dangerous territory. Brunfels, at least, warned his readers when some detail about a herb was debatable.

In his *Iudicium* for the herb *Sennicula* (also called *Sanicula* or *Diapensia*), known in German as *Sanictel*, Brunfels discussed his inability to find sufficient information in the authorities. Without a stable, trustworthy source, Brunfels implied the need for haruspicy and auspices, both forms of divination, in order to make sense of the jumble of information, a practice that lead him to leave behind even the nomenclature of the authorities. At this point, Brunfels found it appropriate to express a firm opinion, namely, that he agreed with herbalists that *Sanicula* was in fact *Solidago minor* and he was so certain of this that he recommended discontinuing the use of the Greek nomenclature and abandoning whatever names are found in Dioscorides and Pliny.[71] Brunfels was not only relying on his own knowledge of and experience with this herb, but he had sufficient confidence to finally make a recommendation about its nomenclature.[72] For the herb *Buglossa*, some sources called it *Cirsion*, which was equated with *Borago* by Dioscorides, by reference to Marcus Cato.[73] Pliny himself claimed that while the leaves of *Cirsion* were similar to those of *Buglossa*, they were smaller and an off-white color, stating simply that *Cirsion* is a tender, little sprout (*Cirsion cauliculus est*).[74] In his 1493 *Castigationes*, Barbaro provided a correction for this passage in Pliny:

> (*Circesion cauliculus est*). This should read either *Cirsion* or *Crision*, per Dioscorides, since it is helpful for varicose veins, which Hippocrates calls "kirsas" and Aristotle calls "ixias."[75]

From this, it is clear that the edition or manuscripts of Pliny to which Barbaro had access did in fact read *Circesion* instead of *Cirsion* and this emendation has been adopted by modern editors. Barbaro elaborated in his *Corollary* on Dioscorides on the ubiquity of the herb, claiming that if it is not *Cirsion*, that would be remarkable, since it would be unnamed (*indicta*) for the ancients, when in fact it is extremely well known (*vulgatissima*). He did allow that it could be a kind of *Buglossa*.[76] Brunfels summed all of this up in his *iudicium*, in which he told his reader that the decision was not his, but theirs, based on the image he painted for them of each plant, *Buglossa* and *Borago*. The reader was also provided with the words of Dioscorides and other learned individuals, so Brunfels felt comfortable in stating his opinion that the two plants are the same in several respects, in particular their flowers, leaves, and medicinal properties.[77] This marks the first time that Brunfels directly addressed the reader, and the first thing he did was place the onus for making a decision about the competing authoritative accounts onto them. Brunfels did his duty by describing both *Buglossa* and *Borago* and providing the opinions of esteemed authorities.

Adding to all of this confusion is Brunfels' discussion of another plant he called *Narcissus*, which is not *Merzenblümen*. He declared that he had literally been tortured over the name of this flower for almost two years.[78] Yet he still could not verify it nor could he wonder enough how this herb, which was so well known and such a common flower, had somehow or other avoided description by any of the authorities. In this case, Brunfels was referring to one of the few situations about which he had a very strong opinion: the source of one's knowledge of a plant. He found it miraculous that this herb[79] was the source of so much confusion, when it was so well known to German medics and herbalists, a perfect example of the difference between book learning and experience. Brunfels chose to believe what the herbalists said about *Narcissus*, since they gained their wisdom from practice and experience of things, not books (*experientia rerum edocti erant*).[80] Yet, despite his advocacy for empirical observation, Brunfels could not bring himself to assert the truth or validity of what he learned. Instead, he placed these phrases and terms in clauses that subverted them, saying "I *might* have understood" not "I understood." Brunfels then went into far more detail than was typical for an *iudicium* about the plant's growth habits, climate, and physical characteristics. From all of this information, it can be inferred that Brunfels was particularly familiar with this plant and likely had experience with both collecting and propagating it. He described its root in terms of direct observation and experience: "I have observed it" (*observavimus*), and "I have experience with it" (*experti sumus*).[81] Due to his own extensive personal knowledge of this plant, he took the liberty of bestowing the name of *Narcissus* upon it, concluding by ensuring that his reader knew that his *iudicium* was backed up by the authorities, but advising them to make their own decisions.[82]

Brunfels did sometimes overtly express wonder in his *iudicia*, for example, in *Pes corvi*, or *Silago*,[83] where the term occupies the first space: *Mira est*.[84] But this is not a true wonder. It is not actually aimed at the fact that there is disagreement among the authorities. Rather, Brunfels was experiencing a growing disbelief in the fact that he could not simply rely on the authorities, but must turn inward towards the reasoning process, based on observation and critical thinking.[85] Further evidence of this turn to reasoning and observation is found in Brunfels' *iudicia* for the two herbs *Cynoglossa* and *Hepatica*,[86] where the verb *comperio* (to ascertain) is used, a term that indicates verification and discovery. *Comperio* is rare for Brunfels, who suddenly decided to mention the importance of using his illustrations for verification purposes. Regarding *Hepatica*, he noted that according to Hieronymus Haerbarius, there were three species and while he was putting together his herbal, he had specifically requested images to be made of the first two, meaning that he already knew of them. The third had been presented by reference to the authorities. The images provided correspond to the first and third and nicely agree with the descriptions of Ermolao Barbaro and Hieronymus Haerbarius, a very early example of verifying the account given of a plant by means of an image drawn from life. The importance of this technique seems to have passed Brunfels by, however, at least up to this point. Most of his verification processes until now had been based simply on prior experience whereas the usefulness of the images was relegated to the process of collecting, but not scientifically identifying, plants in the wild.

In the case of *Capillus veneris*, Brunfels again addressed the reader, but this time, he seems to have gained some confidence. He explained how the images he provided related to the description of the plant found in the ancients, versus the contradictory account of the much-maligned *officinae* (apothecaries) and "neoterics," who considered it *Adianthon*.[87] By "neoterics," Brunfels generally meant either the author of the *Pandectae*, or those who were heavily influenced by the text. Though the *Pandectae* are dated to *c.* 1317, a few printed editions were published in the fifteenth century, with the most recent for Brunfels being the 1488 *Liber Pandectarum Medicinae*.[88] In this edition, the entry for *Capillus veneris* is chapter xcix, which begins with various known names. This being a German edition,[89] the first name provided is *Berscegarten persice*, followed by various Latin names, including *Capillus porcinus* and *Capillus veneris*. The "author" of the *Pandectae* simply collated accounts of the herb from other sources, including Serapion[90] and Dioscorides. The entry to which Brunfels referred is from the former, who equated *Capillus veneris*, *Adionton*, and *Politricon*.[91] That the *Pandectae* are mostly unoriginal and cite ancient authorities provides a certain precision to Brunfels' complaint that they had really mixed everything up.[92] The issue is not so much that the *Pandectae* presented incorrect medical information, although that is certainly a problem, but that it was doing so while citing ancient authorities. The transmission of these texts through the *Pandectae* and the general public's knowledge of them suffered in tandem with those who took its practical, medicinal advice. It was much safer to listen to the opinions of *haerbarii* (herbalists), *vetulae* (old women), and *olitores* (gardeners), all of whom had experience with these herbs and did not need to consult compendia of questionable veracity. Brunfels rejected the "neoteric" take on *Capillus veneris*, which he claimed was a type of *Saxifrage*, as Pliny categorized it, a garden herb so common that it was essentially a weed that got tossed out. Regarding the illustrations, Brunfels confessed that he had made do with images of the common species whenever a particular species was not available for reference, exhibiting an astounding blindness to the fact that Weiditz's illustrations have value as tools for identification.[93] It also completely undermines all of the work Brunfels had done to tease out the confusion surrounding the nomenclature of the herbs. Yet these criticisms, justified as they may be from the standpoint of the history of botany as a science, are not the sum of his work. As these analyses of Brunfels' *herbae nudae* and *iudicia nostra* demonstrate, Brunfels' original Latin began to develop as he was forced to confront the fact that he had no sources upon which he could draw. The features of this Latin seem unremarkable on their own, but in contrast with Pliny's natural-language Latinity, the incantatory Latin of medieval herbals, and the formal botanical Latin that would later develop, they come into focus as an important linguistic transition.

Conclusion

The Italian humanists who critiqued Pliny were all dealing with him at a remove, in terms of his text, its errors and idiosyncrasies, and not in terms of the substance of the work. This is true even for the argument between Leoniceno and Politian, who were engaged in textual criticism, not scientific verification of the claims made in Pliny. They did not find the plants in question, visually examine them, and compare this analysis to the text. There is a simple reason for this. Not one of the Italian humanists who wrote about Pliny's text were herbalists, skilled or trained in medicine and natural history. In light of this, two questions arise: what exactly were they criticizing and how could they possibly have claimed to be correct themselves in their emendations? By focusing so much energy on establishing an authoritative version of Pliny's text, the Italian humanists were faced with the consequences of the long, corrupted manuscript tradition. The humanist concern for the real consequences of incorrect plant identification and description (for example, accidental poisoning) is evidence that they were trying to assuage their anxiety about the ambiguity and incompleteness that had long characterized the florilegia and herbal traditions. This frustration on the humanists' part took partial form in their own appropriation of the terminology of medicine. They focused on "healing" Pliny's text, which was a pathological issue for them. They stressed how mistakes in the text could be fatal, but they also placed a moral value on a correct, authentic, well-collated and edited text, regardless of any question of Pliny's own motives and methods. Where Pliny had criticized Asclepiades' methods and motives, the Italian humanists questioned Pliny himself. By focusing on the stabilization of Pliny's text, however, his emenders missed a key point: Pliny had synthesized the methodologies of Theophrastus and Asclepiades, a fact that sheds light on the import of his rhetorical asides in Books 25 through 30. As a result, the stabilization of Pliny's text was also a stabilization of Theophrastus' and Asclepiades' texts as objects of reception. Further, Pliny recognized the importance of the connection between nomenclature and discovery, implying that a plant with an inferior name that does not provide information for identification purposes, is *inventa* (unknown).

For these reasons, Pliny's text is an invaluable object of reception and transformation. In the same way, the reception of his text in the late fifteenth century is a resource that can be mined for clues regarding the development of botany in the sixteenth century. Not only did the humanists assist in stabilizing Pliny's text, but, as a result of their own input on the subject matter, they also produced a body of texts supplementary to Pliny's, which contributed to a new discourse surrounding the ancient author. It was within and because of this discourse that the first botanical texts of the sixteenth century appeared. This positive take on the inherent value of Pliny's text and its reception was long ignored by historians of botany, science, and the Renaissance. Brunfels is relegated to a mere footnote in Morton (1981), who argued that it was the translation of Theophrastus into Latin in 1483 that precipitated modern botany, overlooking the popularity and influence of Pliny that culminated in the *editio princeps* in 1469. Morton is aware of the fact that in the course of trying to map the plants that grew in their environment to antiquity and the medieval herbals, early botanists were forced to acknowledge that some plants could not be so identified.[94] However, he seems unaware that Brunfels had encountered this issue in 1530. Rather, Morton cites Antonio Musa Brasavola's 1536 *Examen Omnium Simplicium Medicamentorum* as the first botanical text to remark on this problem, which would be correct if self-awareness in early botanical authors were the main criterion. Brasavola remarked that Dioscorides and Pliny did not describe even 1/100 of the herbs in the world, whereas Brunfels simply expressed frustration at not finding the *herbae nudae* in his sources. However, from the standpoint of attempting to describe, name, and categorize these unknown plants, Brunfels' efforts led directly to identification efforts by his successors.

Brunfels' herbal is, as a liminal text, important for the eventual formalization of botanical Latin and the stabilization of botanical nomenclature, which the ICN explicitly credits to Carl Linnaeus.[95] Yet the fundamentals of modern botanical Latin pre-date Linnaeus and can be seen in Brunfels' Latin, which makes use of binomials, paratactic ablative constructions, and increasingly technical vocabulary. As a transitional text, however, Brunfels' herbal displays some key differences both between scientific botanical Latin and the herbal tradition and antiquity, especially the focus on healing. As far back as Theophrastus and Pliny, descriptions of plants were dominated by two main criteria: physical descriptions of roots, leaves, and flowers (not necessarily drawn from personal observation), and details regarding their medical uses. The move away from the latter had not yet solidified by 1530, nor even by 1542 with Leonhard Fuchs' *De Historia Stirpium*. Both writers continued to discuss the medicinal properties of plants. The seventeenth-century botanist Kaspar Bauhin (1560–1624), who was not a physician, also considered a plant's medicinal properties to be innate and in need of scientific description.[96] The disappearance of medicine from the field of botany was gradual, yet even so, Bauhin's text is quite late for such a stance, especially when botanists like Carolus Clusius (1526–1609) had already by 1601 turned away from medicine in their botanical works. Nor was this trend simply an intellectual decision about content. The move away from medicinal descriptions meant that the vocabulary of botanical Latin evolved to include precise terminology for a plant's physical features and to exclude terminology for medical practices. In Pliny, the Latin has an easy simplicity associated with a natural language, whereas in Leonhard Fuchs (1542), it has an economy of expression that is deliberate and precise because the goal is to be intelligible. The botanists of the sixteenth and seventeenth centuries grew increasingly concerned with resolving the issue of proper plant identification, a concern that was reflected in the stabilization and formalization of nomenclature as well as in the development of a laser focus on the plant itself, rather than on its uses. As a result, there was a need for more and more technical terms for the increasing knowledge of the parts of a plant, and less need for a complex medical lexicon.

Brunfels' style lies somewhere in between. The beginnings of this shift can be seen in his vocabulary. He referred to the healing properties of herbs over and over again, but rarely used the verb *sano* (to heal), which is prevalent in Pliny. Rather, he used *praestat* (to be helpful), *curo* (to cure), and *valeo* (to be efficacious). This is remarkable in part because his predecessor and inspiration, the humanist

Ermolao Barbaro, had in 1493 used the term *sano* to refer to his own emendations of Pliny.[97] There had already been a shift in the vocabulary of "herbals" away from terminology associated with medicine, and toward more strictly botanical terms. Brunfels' use of *sano* was not anachronistic; he used it in similar contexts to Barbaro. The appropriation of *sano* by Barbaro for the stabilization of an authoritative text was carried over to the context of botany by Brunfels. There is also in Brunfels a general tone of anxiety and frustration, especially with regard to the *herbae nudae*, which prevented him from stabilizing his *cognitio* about herbs. Brunfels refused to paint himself as an authority, relying instead on others and demonstrating an overall reluctant, even lackluster tone when forced to provide his own opinion or observations, whilst simultaneously promoting his own experience over book learning. He also employed three distinct techniques for the identification of herbs: comparison with other plants, especially by reference to leaves, roots, and flowers; discussion about nomenclature, and the linking of various names to one another; and useful, lifelike illustrations.

Brunfels' Latin was also very much influenced by both the medieval herbal tradition and by the reception of Pliny in the fifteenth century. Stearn (1966) claims that classical Latin was not suited to the task of describing plants in the manner that Renaissance-era science needed. Its vocabulary was sufficient, but knowledge of plants was limited to medicinal uses, habitats, and growth habits. As such, classical "botanical" Latin does not delve deep enough into plant morphology to warrant more technical vocabulary in order to indulge in detailed comparisons. "That is essentially the task of a philosopher rather than a herbalist."[98] For Stearn, the form and the vocabulary of modern botanical Latin both have an origin in a fundamental change in the way botany and the study of plants was conceived. Similarly, Brunfels' method of comparison was more philosophical than the standard medieval herbalists' approach. If so, this is yet another way in which his work is important for the botanical Latin that came later. If we grant Brunfels a degree of credit for the development of botanical Latin, then its origins can be said to reside in confusion and chaos, which are the result of Brunfels' somewhat misguided attempt to imitate his predecessors. This helps to explain his growing disbelief in the wisdom of Pliny and other authorities, as if the herbal is a diary recording Brunfels' developing apostasy.[99] This chaos and anxiety is also explained in terms of transformation methodology, with the caveat that Brunfels was interacting with a reference sphere without realizing that what he was engaged in was a form of reception. He tried to reconstruct the authoritative stances on the herbs that he could not accurately identify, but in the process, was forced to provide his own opinions and experiences, thus appropriating the reception sphere itself and transforming it into a new product, a proto-botanical text in which can be seen the markers of what would eventually become formal botanical Latin.

Many historians of botany have dismissed Brunfels under the assumption that he regurgitated the writings of his predecessors. Yet the ways in which Linnaeus and his contemporaries thought and wrote about plants were the outcome of a slow, simmering progression from the Latinity of Brunfels. It does not follow from the fact that Brunfels' work was superseded by his successors that it has no bearing on the development of the discipline of botany. The ways in which Latin was used in the discipline were codified over time, until in the twentieth century an international governing body instituted rules based on the turns of phrase, syntactical constructions, and vocabulary already employed by the community of botanists in Europe. The mere fact that Brunfels contributed to the discipline, and did so in the Latin language, is significant. Broadly speaking, therefore, there are two categories into which Brunfels' contributions to the development of botanical Latin fall. First, by writing in Latin about German plants (or plants known in Germany, with German names), he showed fellow German physicians and scholars that the language in which the Italian humanists wrote, communicated, and engaged in debate was accessible to them. Moreover, he demonstrated the real-life implications of the philological disputes in which the Italian humanists were engaged. The stabilization of Pliny's text over the course of several printed editions, which culminated in the Barbaro-influenced 1497 edition, was significant for a physician like Brunfels, living and working in Germany. Brunfels himself, through experience, already knew of the medicinal benefits of the plants included in his herbal, but he recognized the need to lend credence and authority to his entries by referencing ancient and medieval authorities and by writing in a language long associated with classical scholarship.

Brunfels' second category of influence involves his overall methodology. His main goal was to verify his *cognitio* (knowledge) of medicinal herbs by mapping their German names onto Greek and Latin names. He began with the idea that his own knowledge of these plants was not sufficient. Confronted with certain herbs that he planned to include in his herbal, he realized that his own experience with them was valuable information. This prompted him to hesitantly describe them, which in turn induced Fuchs to attempt to follow up on them. Brunfels' declaration that others could deal with the intricacies of plant identification (*alii viderint*) began as a shifting of responsibility and a self-positioning as unqualified to make final judgment calls. Even in his own volume, however, this attitude eventually transmuted into a reluctant acknowledgment of the value of his own experience and observations. Where these could not be relied upon, Brunfels settled on asking his readers to make their own decisions. In this way, Brunfels' herbal functioned as an *allelopoietic* supplementation: in the case of certain herbs, he could not simply reconstruct the opinions of the authorities and therefore had to insert his own *iudicium* for his readers. This crowd-sourcing can also be found in Linnaeus, in whose work classification and taxonomy became a communal effort. In Brunfels, the rationale behind this appeal to his readership was still informed by the moral component of his work. As a physician, he saw the accuracy of his herbal as a moral issue. He could not be responsible for misinformation that led to the harm of a patient. In Linnaeus, the confusion of plant nomenclature and its attenuating issues of correct identification and description had simply become a scientific problem to be resolved.

3

Botany from Leonhard Fuchs to Linnaeus and Beyond

In the sixteenth century, the field of botany and botanical Latin developed in specific ways that were a result of Otto Brunfels' 1530 herbal, *Herbarum Vivae Eicones*. By verifying and mapping plants with German names to the Greek and Roman traditions, Brunfels set the stage for later herbalists and botanists to do the work of classification of unverifiable plants (*herbae nudae*), by providing rough descriptions and the realistic illustrations of Hans Weiditz. As a reception sphere, Brunfels looks back to Pliny, but as a reference sphere, he also looks ahead to Leonhard Fuchs' 1542 herbal, *De Historia Stirpium* (An Inquiry into Plants). Because the *allelopoietic* process is reciprocal, one can always, from any point on the historical continuum of botany as a discipline, choose to look forward or backward. Both options, of looking further back or farther forward, are entirely compatible with transformation methodology. The various *allelopoietic* loci that make up a transformation are all points on a three-dimensional continuum that looks simultaneously forward and backward, up and down, side to side, inside and out. From Pliny, it is possible to continue the backwards regression and further investigate his Greek, Egyptian, Persian, and other influences. From Brunfels, the sixteenth-century botanical texts that proliferated can be taken collectively as a reference sphere for his herbal. Beginning from Leonhard Fuchs, who directly cited Brunfels, it is possible to continue the forward momentum and identify all of the botanists whose works were influenced by him.

The links to Linnaeus from Fuchs are easy to establish, because Linnaeus was careful to cite the names and taxonomies of authors who preceded him, including Fuchs. The jump from Fuchs' Latinity in the sixteenth century to the highly formal botanical Latin of the nineteenth century, however, is daunting, characterized by innumerable small, sometimes seemingly unimportant changes, all of which resulted in a highly regulated, formulaic genre of scientific literature that, while using the basic syntactic features of the Latin language, is so divorced from the literature of antiquity as to seem unintelligible. From a methodological standpoint, it makes sense to treat *allelopoietic* points on the continuum of botanical Latin as case studies, examples of botanical texts in a particular time and place, but not monolithic. Between the sixteenth and nineteenth centuries, trends in botanical composition and nomenclature formalized while the vocabulary became more specialized and specific. New terminology developed in response to the discovery of more plant and flower parts and certain noun cases began to be used in ways that were unique to botany. Moreover, there was a turn toward descriptions of previously unknown plants. In order to reflect these changes, a small, curated selection of botanical texts is sufficient: Leonhard Fuch's 1542 *De Historia Stirpium*, Carolus Clusius' 1601 *Rariorum Plantarum Historia* (An Investigation into Some Quite Uncommon Plants), Kaspar Bauhin's 1623 *Pinax Theatri Botanici* (Illustration of the Botanical Theatre), Henry Cranke Andrews' 1797 *Botanist's Repository for New and Rare Plants*, and Edmond Boissier's 1838 *Elenchus Plantarum Novarum* (List of New Plants). In Andrews and Boissier, whose works post-date Linnaeus' system of classification and methodology, the turn to formal diagnoses and descriptions is particularly noticeable. Linnaeus did not focus on discovering new plants, but rather changed how they were classified, developing a form of nomenclature called a binomial epithet, in which the genus and species name is followed by a diagnostic phrase of no more than twelve words. This phrase was eventually dropped from the name and expanded into a longer section known as the diagnosis, which Boissier adopted, while Andrews retained almost exactly the same style and format for diagnoses as Linnaeus used, with one significant difference: he translated the Latin for every entry into English.

Botanical Texts in the Sixteenth Century

As one of the earliest printed botanical texts, Brunfels' herbal was influential not only on the field of botany, but on the development of its language of expression, Latin. This point is best made by comparison with how his successor, Leonhard Fuchs (1501–66), handles the same plants that Brunfels either discussed in his herbal or abandoned as hopelessly unclassifiable. This method allows us to identify the precise characteristics of early botanical Latin and the particular ways in which it changed in the decade separating the publication of their two texts. A full comparison of all of the plants in both authors is unnecessary in order to demonstrate this. It will suffice to focus on the three *herbae nudae* (bare plants) in Brunfels: *Kuchenschell*, *Gulden Guntzel*, and *Gauchblům*. Of these, only *Gauchblům* is discussed by Fuchs, the next European "botanist" after Brunfels, rendering his treatment of the herb especially useful for comparison. Moreover, the nomenclature used by Fuchs for this same herb (*Hiberis* and *Nasturtium agreste*) can be found in Andrea Cesalpino's 1583 *De Plantis libri xvi*, as well as in Kaspar Bauhin's 1623 *Pinax Theatri Botanici*.[1] The latter text, despite the many decades that separate it from Fuchs and Cesalpino, is the earliest comparable botanical text to follow them.[2] Cesalpino's and Bauhin's entries under these headings therefore provide further data with which to track the changes in botanical Latin until Linnaeus' 1743 *Genera Plantarum*[3] and 1753 *Species Plantarum*. *Gauchblům* is just one plant, but the various attempts by botanists to identify, illustrate, classify, and describe it demonstrate all of the features of botanical Latin discussed so far. By tracing the changes in its descriptions from Brunfels to Linnaeus, the shift in and development of botanical Latin therefore also become clear.

This method of comparison is straightforward, but one clarification should be made. The botanists of the Early Modern era not only significantly altered the manner in which they described plants, but their vocabulary and nomenclature became more precise and they ceased to describe plants that had already been discovered. When Linnaeus mapped the variety of plant names from antiquity and the Middle Ages onto contemporary, eighteenth-century nomenclature and taxonomy, he did not simply solidify their classification. The specific plants in question also disappeared from the botanical literature. There was simply no further need to describe them, because many botanists, though certainly not all, now knew not only their scientific nomenclature, but also their key diagnostic features.[4] Modern, post-Linnaean botanical Latin generally describes new genera and species,[5] plants that Brunfels would have called *herbae nudae*. But where Brunfels despaired of the task of describing these plants, trained botanists saw such situations as opportunities, not only for the advancement of their own scholarship and careers, but for passing along accurate, "universally" understandable nomenclature and formal diagnoses and descriptions. As a result, the following distinction must be made: in early botanical texts, the specific plants being described generally differ from the specific plants described in botanical texts from the eighteenth century on. Thus, any comparison of the two must use criteria and variables other than the content or substance of the nomenclature, diagnoses, and descriptions. There is, however, significant overlap in the formal descriptions of genera, as opposed to species and varieties within species. For example, the genus *Laurus* was already known in antiquity, even if the particular species described in ancient texts differ from species in Renaissance texts. As is clear with regard to *Gauchblům*, much of the information provided by Fuchs is derived from comparison with a particular genus, rather than other species. Similarly, more attention is therefore paid to Linnaeus' *Genera Plantarum* and comparisons will focus on vocabulary, style, methods of naming, and the categories of description.

Fuchs on Brunfels

Leonhard Fuchs was, like Brunfels, a German physician who authored a herbal in Latin. His *De Historia Stirpium* was published in 1542 and like Brunfels' herbal, featured realistic woodcuts.[6] Perhaps because of such similarities and despite the initial popularity of Brunfels' herbal, Fuchs' text quickly superseded Brunfels' in renown. The perceived authority of Fuchs' herbal is based in part on its reception and his legacy as a comparatively more methodical and scientific writer than Brunfels.[7] Fuchs' methodology and vocabulary were seen as far more precise, which ought to make comparison with Brunfels fairly straightforward. There is, however, an unfortunate gap between the vernacular names used by Brunfels and those used by Fuchs and later botanists. The fact that *Kuchenschell* and *Gulden Guntzel* are unattested in Fuchs, who was also German, shows that he either knew them by different names or used the same ones as Brunfels but did not deem them worthy of inclusion in his own herbal. If so, it is likely that Fuchs also was unable to match them to Greek and Latin plant names and that he had little or nothing to add or alter in Brunfels' descriptions of them.[8]

In order to understand how and why Fuchs described plants, it helps to know how Fuchs viewed the relationship between his work and that of his predecessor, and how he thought he was improving on what Brunfels had done.[9] Fuchs could not ignore Brunfels' influence and therefore dedicated a considerable portion of his prefatory material to him. He called Brunfels erudite and industrious and declared that he was the first to "attempt to assist with and illustrate the field of medical herbs in Latin and thereafter in German writing."[10] He also, however, pointed out some of the mistakes in Brunfels, such as the lack of correspondence between text and image in a few cases. He disparaged the paltry number of plants included in Brunfels' herbal, which were only the ones "in common use." Finally, Fuchs complained that Brunfels used plant names that were either mismatched (not their own) or illegitimate; in other words, he disagreed with several of Brunfels' attempts at identification and naming. Fuchs was diplomatic in his stance on Brunfels, albeit somewhat condescending. He acknowledged the impact of the printer's interference and incompetence on Brunfels' text, which he called "annoyances" (*molestias*). Yet he twice stated that these faults deserved pardon (*veniam merentur*), a possible reference to Politian's assessment of the mistakes in Pliny as being potentially harmful. As a physician, Brunfels should have known the dangers in misidentifying and misrepresenting medicinal herbs. It is almost begrudgingly that Fuchs admitted that despite these flaws, Brunfels was deserving of praise because he recognized the importance of providing accurate and realistic images of the plants alongside their textual descriptions. He claimed that Brunfels was the first of all people to convey anew into Germany the "correct method of illustrating plants" and that by doing so, he also provided to others the chance to imitate him, to produce better quality and more scientific herbals with realistic illustrations.[11]

Fuchs' acknowledgment of the historical importance of Brunfels' method of including lifelike images of plants in their descriptions is precisely the feature for which Brunfels would be remembered, although Fuchs glossed over the fact that Brunfels was not at first convinced of the efficacy of Weiditz's illustrations. It was the publisher Schott's idea and Fuchs must have realized this, both because Brunfels outright stated it and because of the ongoing lawsuit between Schott and his competitor, Christian Egenolph, whom the former accused of plagiarizing the illustrations. It is difficult to say definitively whether Fuchs' statement constitutes a dismissal. Certainly, his praise of Brunfels for this development was well-founded, but it was undermined by the implication that the use of realistic illustrations, which Brunfels did not wholly appreciate, was the only truly remarkable feature of the text. More pertinent is the description Fuchs provided of Brunfels' subject matter, the "medicine of herbs," which had been nearly extinct in Germany until he dragged it out of the shadows (*eruere tenebris attentaverit*). The latter part of this statement is a reiteration of Brunfels' own claim that he brought the field of medicinal herbs into the light. But the claim that it was nearly extinct is unclear. There were many herbals produced in the late Middle Ages and early Renaissance, some in German, others in Latin. Fuchs' statement seems therefore to be directed at the German people, for whom this knowledge had been inaccessible in this specific format: German medicinal herbs described in Latin. As a result, Fuchs thought that Brunfels was "most useful to posterity" and that his efforts in the field of medicinal herbs should be reproduced in all other fields of study.[12] Fuchs acknowledged the inventiveness of Brunfels' herbal and valued his methods, but was careful to describe his subject matter as medicinal, not scientific. This positive estimation of Brunfels' originality and influence unfortunately did not extend historically past Fuchs, who borrowed extensively from Brunfels, extending and improving on many of his ideas. But where Fuchs is given the benefit of the doubt, Brunfels is usually shunted to the side and dismissed as unoriginal.

As the rationale for his own text, Fuchs cited the pursuit of knowledge derived from the ancients. The reason he produced his own herbal was that the subject matter (plants) had since antiquity been considered not just pleasing and useful, but actually worthy of study. Parroting Pliny in Book 25, he said that even in antiquity plants were considered an ancient object of study, endowed with considerable authority and veneration.[13] Fuchs' dedication is formulaic out of necessity, but this appeal to ancient authority and the use of religious terminology to describe it is typical of the tension in earlier botanists, between needing to justify their work to Christian audiences and sponsors, and wishing to establish the study of plants as knowledge-producing. It is also very similar to Brunfels' own *ratio*, of recalling the herbal tradition from near extinction for the benefit of humankind. Given the similarities between the two texts and their near synchronicity, the implication is clear: Fuchs saw his own work as an improvement on Brunfels and therefore even more useful to posterity.

Gauchblům (*Nasturtium agreste*)

While discussing the *herbae nudae* and his frustration at not being able to identify them in the ancient sources, Brunfels had declared, *alii viderint* (let others see to it).[14] In fact, Fuchs did follow up on *Gauchblům*, which he identified as *Nasturtium agreste* (also known as *Hiberis*).[15] In Greek, it was called either *cardamantice* or *agriocardamum*. Similar to Pliny, Fuchs provided a brief etymology of the Greek name *agriocardamum*, namely, that its odor, leaves, and flavor are all similar to either *Nasturtium aquaticum* (which Dioscorides called *Sysimbrium cardamine*), or to *Hortensis*, which the Greeks called *Cardamon*. The entry runs over three pages, in much the same style as Brunfels. The first page in Brunfels, however, shows the relevant illustration, then lists the nomenclature. In Fuchs, the first page starts with the title *De Hiberide*, which indicates that *Hiberis* is the name Fuchs formally selected to signify the plant in question. The first part of the full entry is *Nomina* (Names) and the first line refers to it as *Hiberis*, followed by the Greek nomenclature, then the Latin *Nasturtium agreste*. However, and this is crucial, Fuchs included the ablative *Hiberide* as a qualifier to *Nasturtium agreste* and added the statement *officinis nostris incognita* (unknown in our apothecaries). This means that the plant name *Nasturtium agreste* was not known in Germany, which explains Brunfels' inability to map the German name to a Greek or Latin one. Yet even the name *Hiberis* does not show up in Brunfels, who was only aware of the German vernacular *Gauchblům*. Fuchs acknowledged this, stating that *Gauchblům* must be preferable to Germans, for the simple reason that many people were not yet aware of its renown.[16] Fuchs therefore saw the reasoning behind the retention of this name in Germany as nothing more than attachment due to ignorance. Those who used it only did so because they liked it, and they liked it because they had not yet "arrived" (*pervenerunt*) at acquaintance with either the herb or its broader reputation. They did not know enough about it to recognize it as *Hiberis* or any of the other names ascribed to it. In their ignorance, they promulgated the Germanic name as though it were a different plant. But there was more to Fuchs' reasoning. He explained the etymology of its various names, noting that *Gauchblům* could also be called *wilder Kress* in German, a literal translation of the Greek names *cardamantice* and *agriocardamum*. These two names were given to it for three possible reasons: (a) the scent of its root being the same as that of *Nasturtium*; (b) its leaves being similar both to those of *Nasturtium aquaticum*, called *Sisymbrium cardamine* by Dioscorides, and to those of *Hortensis*, which is called *Cardamon* in Greek; and (c) the leaves resembling *Nasturtium* in flavor. A literal translation of *Gauchblům* is *Gauch* (cuckoo, or *Cuculus canorus*) and *Blum* (flower) and Fuchs evidently thought that this was not descriptive of the actual plant. In fact, if one actually knew what it looked and tasted like, one would compare it to *Cardamon* or to other species of *Nasturtium* and use a German name that reflects this. Fuchs somewhat flippantly suggested *wilder Kress*, which lead him to the physical description (*forma*) lacking in Brunfels. He stated that it had leaves similar to *Nasturtium*, which flourish in spring and are around a cubit in length. The flower is milk-white and supported by two roots. One species of the herb, which has the bitter scent of *Nasturtium*, is more commonly found in Germany. Fuchs then referred to his illustration and claimed that based on this textual description, it should be clear to everyone that the herb depicted in it is *Hiberis*, because everyone knows it has a stem that is a cubit in length and leaves like *Nasturtium*.[17] In this case, Fuchs demonstrated the ablative constructions that would later define botanical Latin: *verno tempore virentia, cubitale longitudine, aut minore, flore lacteo, radicibus duabus*. These are set in apposition to the main verb *habet* (it has), which is in the present indicative, while the phrase *odore Nasturtii quam acerrimo* is subordinate to *nititur* (it is supported), again in the indicative. This move away from the subjunctive reflects an increasing focus on objectivity, but it is also a purposeful rejection of Brunfels' rhetoric of hesitation and anxiety.[18] Fuchs was not a reluctant botanist and he willingly took up Brunfels' charge to do the work he claimed he could not do. Fuchs was not, however, ready to reject Brunfels' method of self-justification by way of referencing the authorities. Fuchs also cited Dioscorides, Galen, and Pliny[19] and included the medicinal properties of the herb, not just its diagnostic features.

The technique of description by comparison with another plant is ancient, and in fact, Fuchs' entries follow a very similar formula to the one used by Brunfels. Imprecision in naming is also an ancient problem, discussed by Pliny himself. In this entry, however, Fuchs did not provide a species name for the other plant called *Nasturtium*. This is either because he did not see the need for specification, or because the attributes he listed are common to all of the species he knew. This is complicated by the fact that Brunfels had an entry in his 1531 Appendix for a plant called simply *Nasturtium*, but he provided a different vernacular name: *Bressen*, not *Gauchblům*.[20] Despite the criticisms Fuchs leveled at Brunfels, he was not any more precise in his nomenclature or any more consistent in his use of the binomial format. For instance, the remainder of the comparative portion of the entry on *Hiberis* shows that Fuchs was in fact aware of another species, called "Garden Nasturtium." In this description, he said that some leaves of the aquatic genus grow straight from the root, but that the leaves that grow at the top of the stalk call to mind the leaves of *Nasturtium hortensis*. They flourish in the spring and also have a milk-white flower with a purplish hue, which is why Dioscorides called the flower *lacteus*. It also produces a seed in its pods, so tiny that, per Pliny, it can scarcely be seen. Finally, its root is sharp and narrow and has the scent of *Nasturtium*.[21] In addition to the heavy use of ablatives, Fuchs deftly combined technical vocabulary (*radix*; *caulis*; *siliqua*) and sophisticated Latin syntax such as subordination and indirect statement ("it is attested that"). The entry also displays almost all of the same features of Brunfels' Latinity identified in the previous chapter. One seemingly minor point of difference involves the source of their identifying information. Brunfels simply described what a particular plant looks like, whereas Fuchs frequently referred to his own illustrations. This tendency on his part hints at his own experience with the herbs in question. As a physician, he may have grown some himself, but he might only have known the form of others after they have been dried or prepared in various concoctions. In that case, he would need the illustrations that he commissioned from skilled artists. On the other hand, he may simply have had such confidence in the quality of the illustrations that he was content to use them for his textual descriptions, or he might have been anticipating his reader's lack of access to the living plant or any specimens of it. Either way, Fuchs was far more aware of the usefulness of his illustrations than Brunfels was and as a result, he actually cited them in his textual descriptions.

Following the sections *Locus*, *Tempus*, and *Temperamentum*, which detail the plant's native regions and growth habits, Fuchs provided its *vires* (medicinal benefits), taken from Dioscorides, then Galen, and then Pliny. He concluded the entry with an appendix, in which he justified his identification of *Gauchblům* with *Hiberis* (*Nasturtium agreste*). Referring to "fairly recent herbalists," meaning Brunfels, he remarked that they recommend using this herb brewed in lixivium to treat lice.[22] However, Fuchs did not explicitly mention Brunfels, despite the fact that he had also specified the distillation of the herb in lixivium. Fuchs asserted that the evidence for *Gauchblům* being *Nasturtium agreste* is that they have the same medicinal benefits. This entire entry is reminiscent of ancient descriptions, but the vocabulary is more technical than anything seen before. There is a noticeable difference between Fuchs' vocabulary and Brunfels.' The latter had a complex medical vocabulary but used more general botanical terminology, settling for, e.g., *folia*, *flores*, and *figura*. Fuchs employed more complex botanical terms and had a separate section for the medicinal uses, called *vires*. Botanical terminology aside, however, and despite Fuchs having advanced the field of the scientific study of plants well beyond Brunfels, there is a major point of convergence and similarity between the two: both Brunfels and Fuchs used terms such as *herbarius* and *medicus* to refer to those who had a professional interest in the identification and description of plants. They both continued to associate the study of plants with medicine. For example, in his entry on *Solidago*, Fuchs said that it heals wounds (*vulnera glutinat*) and eliminates blood clots and bruises (*sanguinis grumos ex casu, vel constusionibus, coactos in corpore deficiit*). Fuchs' vocabulary of healing parallels the increased precision of his entry on the physical characteristics of the plant; *quadrangulus*, *lanuginosus* (woolly), *geniculum* (node; joint), *propago*, and *laciniatus* (fringed) are all far more technical and specific to botany than the terminology of Brunfels. In fact, Fuchs' vocabulary was so new that he provided a glossary of botanical terms before the index of Latin, Greek, and German names at the beginning of the book, a novelty for a botanical text in 1542. In his entry on *Sanicula*,[23] he employed the neoteric botanical terms *dissectum*, *capillamentum*, *fibrata*, *coliculum*, and *subnigris* (nearly black).[24] Other terms in Fuchs are found in classical Latin, but their usage in his text has become far more regulated and formalized. In Fuchs, they mean something very specific. Terms appropriated from classical Latin into botany include *surculus* (sucker shoot), *pusillus* (tiny), *rubens* (ruddy), *ruber* (red), *capitulum* (inflorescence, head), *candidus* (bright white), *striatus* (striped), and *oblongus* (oblong). Eventually, in 1917 and 1966 respectively, Helen Choate and William Stearn would take the time to delineate the botanical meanings of such terms and render them in English. But as is clear from Fuchs, the development of a botanical Latin vocabulary, in which the terms have uses that are not obvious to non-practitioners, had already begun by the sixteenth century.

The Seventeenth Century

In the 1583 *De Plantis Libri xvi* of Andrea Cesalpino (*c*. 1524–1603), there is also an entry for *Nasturtium*.[25] Cesalpino's text diverged significantly from Brunfels and Fuchs. He did not provide any vernacular or alternative Greek and Latin names in his index. In addition, his work is philosophical and as a result, at the beginning of each book he provided a short explanation of how the plants were classified scientifically. For example, all of the plants in Book 8 fulfilled the following criteria: they had two pericarps (follicles, *conceptacula*), with solitary seeds in individual pits (*singulis alveolis*). Their flower was situated at the very top of the fruit, like the genus *Ferulaceum*. Alternatively, the fruit was without a flower (*sine flore*), or the flower is without fruit (*sine fructu*), like *Mercurialis Xanthii*. They did not bear fruit in the umbels (*umbellis*) like *Ferulacea*, but either scattered in the hollows (*alis*) of the leaves and at the very tips of the branches (*summis ramulis*), or on the heads (*capitulis*), like *Pimpinella*.[26] We can divine from this that Cesalpino classified the genus *Nasturtium* according to its flower parts, seeds, and growth habits. The full entry for *Nasturtium* begins with a description of the plant's physical characteristics, with only one name in the vernacular provided. This name is in fact simply an orthographical variant: *Mastorsium* versus *Nasturtium*. It is grown, he declared, in gardens as a vegetable and when raw, like *Eruca* (lettuce), it has a sharp flavor and a stem with many branches.[27] In the 40 years since Fuchs, the language of botany had progressed to the point that Cesalpino's entry on the same plant, although it employs the same method of comparison with other plants, reads as far more technical. The vocabulary (*conceptaculum*; *alveolus*; *ramulus*; *capitulum*) is not only more specific to botany, as opposed to medicine, but shows a trend of diminution. Flowers are generally small in stature and their parts even more so. As a result, terms from classical Latin are furnished with diminutive suffixes to reflect the size of their new referents. This also is evidence of the advancements being made in scientific equipment. The closer the botanist can look at a plant specimen, the more details they can see and the more previously unknown parts need to be named. However, in solidarity with both Fuchs and Brunfels, Cesalpino was still tied to non-diagnostic features of the plant: a modern botanist does not need to know that *Nasturtium* is a foodstuff or that it is bitter to the taste.

The rapidity of the changes in botanical texts was clear by the turn of the seventeenth century. Carolus Clusius (1526–1609),[28] a botanist and the first prefect of Leiden's *Hortus Academicus*, published in 1601 a volume on Spanish and Austrian flora, called *Rariorum Plantarum Historia* (An Investigation into Some Quite Uncommon Plants). The entries are lengthy, similar to those of Brunfels and Fuchs, but with an entirely different focus. For the tree *Siliqua silvestris*, he provided a short explanation of its etymology immediately after the name, in direct contrast to the herbal tradition of listing various known aliases in Greek, Latin, and the vernacular. Clusius noted instead that he imparted (*indidi*) the name *Siliqua sylvestris* to this tree not because it is similar to a peapod described by the ancients (*siliqua* is Latin for a peapod), but because this is its common name (*vulgari nomine sic appelletur*) in some places of Spain.[29] Clusius' reason for the name of the plant is not scientific but a formalization of its common name. Following this statement is a paragraph describing in surprisingly technical terms the defining characteristics of the tree. The first sentence takes the place of what in a modern description would be a diagnosis, stating the essential characteristics of the species that make it stand apart from others of the same genus. Clusius chose a simple point system to elucidate the characteristics of the tree he was describing, stating first how and where it grows and noting that when cultivated, it tends to grow into a tree of reasonable size (*crescit interdum in arborem iustae magnitudinis locis cultis*). In its natural habitat, however, where it is neglected and grows freely, it is more like a shrub than a tree (*sed neglecta et sponte nascens, plerumque inter frutices potius, quam arbores censenda est*). Clusius then described in detail its general growth habit, the formation of the branches, the color of the bark and of the flowers, the texture of the leaves, the shape of the seed, and so on. The terminology is far more technical and precise than in earlier botanical texts. For example, his explanation of the branches, flowers, and leaves is highly specific: it has rather scattered branches, which grow alternately from the sides, and which are covered with bark of purplish-black (*rariores fert ramos, alternatim ex lateribus nascentes, cortice ex purpura nigricante tacto*). It flowers in early spring, either before the leaves appear or simultaneously, in groups of three or four. The flower clusters, which are a handsome purple color, are joined together at their base, similar in appearance to those of heather or of peas. The leaves emerge after the flowers at intervals and are very similar to those of *Asarum*.

The Latin that Clusius has used to describe these features is markedly different from that of the sixteenth-century herbalists. He employed a series of paratactic ablative constructions (placed one after another with no subordination) in order to delineate them, such as "in early spring" (*primo Vere*); "with these (the leaves) just beginning to germinate" (*iis primum germinare incipientibus*); "at the base" (*inferiore parte*); "with a fine purple color" (*colore purpureo eleganti*); and "at intervals" (*ex intervallis*). Clusius retained his most technical phrasing, however, for the leaves, which are described as "less carnosus (fleshy)" (*minus tamen carnosa*) than those of *Asarum*; "of a nearly circinate rotundity" (*circinatae propemodum rotonditatis*); "sinewy" (*nervosa*); hardy (*dura*); green on the upper surface (*superiore parte viridia*); and whitish on the lower epidermis, or bottom (*inferiore vero nonnihil candicantia*). Similarly, the seed pods are rendered with far more precision than one might expect for this time period. The flowers are followed by "purple, semi-transparent, membranaceous pods" (*floribus succedunt membranaceae purpurascentes siliquae*), which are pressed together (*compressae*). Syntactically, this string of ablatives is key to understanding his description. In Latin, the ablative case is like a junk drawer into which one tosses every random syntactical category not covered by the other cases, and its adoption by botanists is not happenstance. It reflects the incredible descriptive capabilities of that case compared with others and at the same time, increases efficiency of expression. It is possible, using the ablative, to state in only a few words what would take an entire sentence in another language, or with finite verbs in complex hypotactic (subordinate) constructions in Latin. It also allows for less punctuation. The meaning of a statement is derived entirely from the syntax and from the noun endings, which are the only way to make sense of the phrases. The order in which the parts are described is also not a coincidence. Clusius refrained from overtly describing the roots of *Siliqua sylvestris*, doing so indirectly by explaining that it grows into a tree in certain conditions, but is bushier when restricted in the wild. Generally, Clusius moved from bottom to top, upward and outward, listing ever smaller parts of the tree, beginning with the branches, then the bark, the leaves, the flowers, the seed, and finally the seed pod, an order adopted by many botanists. We are not told what height the tree achieves and the level of detail provided appears to end with the description of the seed.

Following this precise and economic statement, roughly corresponding to a diagnosis, Clusius provided a far lengthier description with a decidedly narrative structure employing finite verbs with subordinate clauses. The first information he provided was a differentiation of this genus from another, which in modern botany would be expressed in the diagnosis, or in the first paragraph. Here, he stated that there is another genus that is quite similar in all respects to this one, apart from the flower, which is pure white with purple only at the very tip, and except for the bark of the young shoots, also pure white rather than purple. After this comparison, Clusius returned to describing *Siliqua sylvestris* in more detail, stating that it grows freely (*sponte nascitur*) between hedges in the "kingdom of Granada" and some other places of Spain and of France. It is also maintained in gardens not only in Germany and Belgium, but also in the ancient castles of Spain, not for any agricultural or medicinal reasons, but simply because it is considered charming. He did not observe this second, unnamed genus in the wild (*sponte nascens*), but only sprung up by seed (*semine dumtaxat natum*) in some Belgian gardens. There is some confusion regarding his use of the term *genus*, which refers to a particular level in plant taxonomy, below a family and above a species. Clusius, however, used the term in a less formal way to refer only to a kind of organism, more closely aligning with *species*, since it appears that there are in fact two species of *Siliqua* that he discussed, one with purple flowers (*sylvestris*) and another with white flowers. This informal use of the term *genus* corresponds more to its use in antiquity than in the eighteenth century, post-Linnaeus. Despite this inconsistency, Clusius' methodology is much more scientific and aligned with modern botany than that of his predecessors, who focused on nomenclature, verification, and medicinal uses, rather than on detailed, precise descriptions of the parts of the plant.

Clusius' Latinity differs from that of his predecessors primarily in syntax and vocabulary. In Swiss physician Kaspar Bauhin (1560–1624), there is an even greater stylistic remove from the sixteenth-century physician-herbalists Fuchs and Brunfels. Unlike them, Bauhin considered himself a botanist as well as a physician, even using the term *botanicus* in his 1623 *Pinax Theatri Botanici*. The text also has a separate preface addressed to his readers who identified the same way, called *Ad Lectorem Botanicum* (To the Reader who is Botanist).[30] Nevertheless, Bauhin still aligned himself with the medical field, declaring that any botanist who wants an arsenal of medicaments (*auxiliorum undique copiam*) to draw upon should become an expert in plants, animals, metals, and all terrestrial bodies that are known to have medicinal benefits. By studying these subjects, the botanist can distinguish between true remedies and fake ones, between what is accurate (*exacta*) and what is counterfeit (*notha*).[31] There is a simple reason for this loyalty to medicine. Much like Pliny, Bauhin saw medicinal properties as innate, as part of a plant's nature. In order to know a plant, one must therefore know its uses.

Despite Bauhin's adherence to the principles of medicine, his writing style is strikingly different from that of his predecessors. He refrained from lengthy descriptions of genera and species, employing as few words and characters as possible, anticipating Linnaeus' binomial epithets. His approach was so economical that he even provided at the beginning of his text a register of abbreviations, possibly to lower the printing costs. Despite this, the volume runs to over 500 pages. The point of the *Pinax* was to collate the various names of plants and to clear up the confusion surrounding them, which meant that any description of a plant's features was directly related to its nomenclature. There is, for example, no entry in the *Pinax* for the much discussed and debated *Nasturtium agreste*. However, in the index, he listed the following species of *Nasturtium*: *Aquaticum*, *Babylonicum Lob.*, *hybernum Thal.*, *hortense*, *Indicum*, *Maritimum Lugd.*, and *Montanum*.[32] There is a separate entry called *Nasturt. Palustre Ges.*, with a *Nasturt. Pratense Ges.*, and a *Nasturt. Sylvestre Thal.* It is unclear from the nomenclature whether any of these correspond to *Nasturtium agreste*. Since the index does not provide answers, the main entry in the body of the text is the only way to determine the relationship between the species of *Nasturtium* recognized by Bauhin and *N. agreste*. Under the subheading *Nasturtium Eiusque Species* (*Nasturtium* and its Species),[33] Bauhin identified *Nasturtium* with the *Kardamom* described in Dioscorides[34] and Theophrastus.[35] He claimed that the name *Kardamom* derived from the fact that it "afflicts the head with its own heat and bitterness." On the other hand, it may also derive from the fact that it "produces heart-shapes" (*quod cordatos faciat*); or, it may be because it is useful for heart conditions like "heart skipping" or fainting (*quod in Syncope Cardiaca dicta, plurimum valeat*). The possible etymologies of *Kardamom* have an unspecified relationship to the Latin name *Nasturtium*, which refers to the pain it inflicts on the nostrils (*a naribus torquentis*). Bauhin then discussed the known species (using the term *genus*, as did Clusius), attempting to draw further on the authority of Dioscorides and Theophrastus, despite the fact that the latter only said (4.*hist*.7) that there were many kinds and the former only said that *N. Babylonicum* was the best. Bauhin turned to the Roman tradition instead, to Pliny, who identified *N. hortense* (sown-by-seed), which has broader leaves and a white seed, and *N. sylvestre* (of the woods), which has black seeds.[36] Bauhin further divides the species into *N. hortense* (garden), *N. pratense* (meadow), *N.* a*quaticum* (swamp), and *N. montanum* (mountain).

One of the most striking features of this entry is the increased efficiency of the writing style, which is syncopated in much the same manner as Ermolao Barbaro's *Castigationes*.[37] The syntax is distinct, with so much ellipsis that it is not always obvious to which clause a verb belongs. In addition, Bauhin abbreviated wherever possible, rendering the text inscrutable to anyone unfamiliar with the ancient texts, or the plant names and plant parts. Much like Clusius and other early botanists, Bauhin seemed to use the term *genus* synonymously with *species*. In addition, as the remainder of the section shows, Bauhin did not only use the binomial format, but often employed polynomials. For each of the four categories of *Nasturtium*, he provided a number of subspecies with names such as *Nasturtium pratense magno flore* (meadow Nasturtium with a large flower). This style would later be formalized by Linnaeus, who included in his definition of a binomial epithet a short diagnostic phrase in the ablative, of no more than twelve words. There are no explicit references to medicinal benefits, which is peculiar, because if medicinal properties are essential characteristics of a plant, one would expect Bauhin to list them. It is sufficient for him to mention the possible etymology of the Greek and Latin names, which refer to ailments and side effects. By delineating the physical features of the plant that may have served as inspiration for its name, the botanist has also provided useful information for identification of the plant. By referring, or deferring, to ancient authors who list the medicinal benefits of plants, Bauhin has both absolved himself of the responsibility to do so and saved space in his own text.

Bauhin listed the names of all of the kinds of *Nasturtium* that fit into each of the four categories. For each subspecies, he named the authorities whose descriptions of *Nasturtium* fit with that particular species. For *N. hortense vulgatum*, he added that this corresponds to *N. album* in Pliny, to simple *Nasturtium* in Brunfels, and to *N. hortense* in Fuchs.[38] Bauhin provided a short summary of two subspecies, stating that, "one has simple leaves" (*foliis simplicibus*), and is either smooth and rather white or slightly hairy and less white; and the other has "clearly divided leaves" (*foliis valde divisis*), and is either large or small.[39] *N. hortense crispum* is "both broad and narrow of leaf, and each variety is described and depicted in Matthiolius."[40] The next category is *pratense* and Bauhin aligned the subspecies *N. pratense magno flore* with *Flos cuculi* in Brunfels and with *Hiberis* and *Nasturtium agreste* in Fuchs. Bauhin was primarily linking names to other names and was not necessarily engaged in careful analysis of each species of the plant. *Hiberis* and *Nasturtium agreste* are the same plant in Fuchs, which means it should have been sufficient for Bauhin to link it to the name in Fuchs' index. However, he wanted to be as precise and thorough as possible and since Fuchs used the two names somewhat interchangeably, Bauhin was careful to list them both. Of this species, Bauhin simply stated that its flower is usually light in color and very pale purple, but that sometimes it is straightforwardly white.[41] Of the remaining subspecies of *Nasturtium*, only a few others correspond to plant names in either Brunfels or Fuchs. *N. aquaticum supinum* corresponds to *Sisymbrium Cardamine* in Fuchs; and *N. sylvestre Osyridis folio* corresponds to *Bursa pastoris minor* in Brunfels and to *Thlaspi angustifolia* in Fuchs; *N. sylvestre tenuissime divisum* corresponds to *Nasturtium sylvestre* in Fuchs, but also to *Seriphium absynthium* in Fuchs.[42]

Bauhin's entry for the genus *Pulsatilla* is also useful for making sense of the nomenclatural and taxonomic confusion surrounding one of Brunfels' *herbae nudae*, specifically, the plant he called *Sacketfraut*. Bauhin provided an etymology for the name *Pulsatilla*, which refers to the *pappi* (sepals) trembling at the slightest breeze (*tremuli pappi levissimo flatu*).[43] More importantly, for the species called *Anemone*, he provided links to both Pliny and Theophrastus,[44] which provided context for Brunfels, whose index does not list any of the alternative names provided by Bauhin. Yet Bauhin himself listed the following name:

> *Pulsatilla folio crassiore et maiore flore. Rüchschell/Brunf.*
>
> Pulsatilla, with rather thick leaf and larger flower. *Rüchschell* in Brunfels.

As is typical in early botanical texts, owners would often further cross-reference the names, especially to Linnaeus' classification. In this case, one can find in various physical copies of Bauhin the manuscript marginalium *anem. pulsatilla, L.*, the formal botanical name endorsed by Linnaeus. While Bauhin's use of Latin diverges from that of the sixteenth-century herbalists, his techniques and assumptions do not. Like Brunfels, Bauhin was engaged in a process of verification, trying to make sense of the jumble of Greek and Latin plant names and to identify which ones refer to the same plant, in which authors. Like most of his predecessors, he considered medicine and botany to be related fields.

Linnaeus and the Eighteenth Century

By the eighteenth century, the Latin used to describe plants had continued in its formalization of vocabulary, syntax, and style. However, all traces of medicine and medical terminology had completely disappeared. One of the few concerns of Pliny and the sixteenth-century botanists that remained was the idea that experience led to knowledge. In the second edition of *Genera Plantarum* (1743), Linnaeus stated in his prefatory remarks that he himself, being tried and tested in things (*expertum rebus docuere probatum*), had been taught by hard work and dedication (*labor et studium*), with which he had occupied his days. Linnaeus was addressing the long and winding, utterly confused history of his discipline stating once and for all that any student of botany as a science must, once they have learned from it what they can, turn their gaze away from this history and rely on their own hard work and effort. There is a level of irony in this claim, because Pliny had made a similar argument in favor of experience instead of learning from books, but there is a crucial difference. Linnaeus was intent on looking forward in order to leave behind any plants that are already known, named, and described, as long as this has been done accurately and in accordance with his own methods and criteria. In the future, any plant description was to be novel, not in term of its style or the nature of the contents, but in terms of the plant in question.

One of Linnaeus' most significant contributions to botanical Latin, and perhaps least well-known, involves his formalization of the ablative case in plant names. Earlier botanists had mostly used the binomial format of genus and species, where the latter is usually an adjective matching the former in gender, case, and number. Some, such as Bauhin, had included very short phrases in the ablative, but had done so informally and irregularly, as a method of distinguishing between subspecies. In contrast, Linnaeus saw the ablative as a crucial component of a true binomial, because a species name was not simply adjectival and therefore descriptive, but also diagnostic. It differentiated the plant from others, not just for identification purposes, but also for classification. In Aphorism #257, in both the *Fundamenta Botanica* (1736) and the digest form *Critica Botanica* (1737), he stated that a specific name (*nomen specificum*) distinguishes a plant from all other types. In Aphorism #258, he added that the specific name "will reveal at first glance the plant to which it belongs, because it contains the uniqueness assigned to the plant itself." These two axioms underlie his insistence on very short ablative constructions within the formal diagnosis, which itself was to be no more than twelve words in length and clearly expressed both the differential and the essential characters of a plant.[45] In this way, he described the plant *Bauhinia divaricata* with the phrase *foliis ovatis lobis divaricatis* (with ovate leaves and spreading lobes), so that it was distinguished from *B. ungulata*, for which he used the phrase *foliis ovatis lobis parallelis* (with ovate leaves with parallel lobes), and in contrast to *B. variegata*, described as *foliis cordatis, lobis coadunatis obtusis* (with cordate leaves and blunt lobes that are united at their base).[46] These short phrases were treated as part of the name of the species. However, the parataxis and punctuation used are crucial to deciphering them. For example, Linnaeus used commas to indicate when a particular ablative phrase referred to the plant overall, as opposed to providing further information about the plant part just described. While Linnaeus insisted on incredibly short diagnoses, his successors found it difficult to keep to twelve or fewer words. Over time, these epithets became slightly longer and were separated out into a new diagnostic paragraph, although they still retained the ablative constructions.

Traditionally, as is clear from Clusius, what followed the diagnosis, whether it was extremely short or somewhat longer, was a brief statement about the geographical areas to which the plant was native. Place names were either taken from antiquity, the Middle Ages, or were modern names that had been Latinized.[47] The purpose of such statements was to indicate where a plant may be found and to that end, they commonly take the form of the preposition *in* + ablative, which translates to "in/on," and *ad* + accusative case, which translates to "in the direction of/towards/by." For example, the phrase *in provincia Cajatambo in montibus Cordillera negra ad viam ad oppidum Ocros ad jugum Chonta dictum* translates to "in the province of Cajatambo, in the Cordillera Negra mountains, on the road to the town of Ocros, by the ridge known as Chonta."[48] After the geographical statement, the entry would include the formal description, which, in contrast with the formal diagnosis, was in the nominative case, not the ablative modifying the genus name. The formal description had considerably fewer restrictions on it than a formal diagnosis. It could be much longer and while opinions differed over the years, Stearn (1995) asserts that a description ought to at least provide information on habit, form, and the ways in which the plant differs from others in the same taxonomic groups.[49] Similarly, over time, while the information included in a scientific description differed, more and more botanists were unfamiliar with Latin as a language of expression, and the order of the information became systematized. Thus, for flowering plants, Linnaeus' rule from his 1751 *Philosophia Botanica* still holds. "Let the description follow the order of growth . . . it is best to follow the natural structure from the roots (*radice*), to the stem (*ad caulem*), the petioles (*petiolos*), the peduncles (*pedunculos*), and then to the flowers."[50] In other words, Linnaeus recommended that the sequence of a description move up, in accordance with the growth pattern of the plant. Similarly, Alphonse de Candolle's 1880 recommendation also still holds, to "pass from the known to the unknown, from definite matters to indefinite ones, from those which are most apparent to those which are less so."[51] Taking these two maxims into account, for a description of a plant's leaves, one might state the position on the stem, how many leaves there are, and their overall shape and outline, followed by descriptions of the apex, margin, base, length, breadth, indumentum, veining, texture, and color.[52] In general, the petiole and stipules of a leaf are described after the blade itself, whereas for flowers, the anthers and the pollen follow the description of the stamens. There is a direct benefit to such conventions. If everyone agrees to follow them, those who compose descriptions can compare and contrast various species in a methodical way that makes sense to others in the field. Plants described with similar wording can be taken as similar and vice versa,[53] and the method of comparing and contrasting is strengthened by the deliberate use or avoidance of specific terms. This is further supported by the trend of italicizing or altering the kerning of certain terms within a published description, in order to bring the reader's attention to them. For the modern botanist, the formula of species name + ablative constructions was further detailed so that there was even a recommended usage of semi-colons and commas, which were preferred over periods with their need for capital letters for particularly lengthy diagnoses.[54]

Linnaeus' influence on the field of botany cannot be underestimated, but it must also be noted that this was a pre-existing convention that can be seen in far earlier botanical texts. Linnaeus' authority, however, had a broader, global impact. Where for earlier botanists, such methods were conventional, Linnaeus' adoption and recommendation of those same techniques had the effect of stabilizing them, rendering them *de facto* requirements. In the case of the genus *Nasturtium*, about which there was so much confusion in previous centuries, in 1753, Linnaeus reclassified it as *Cardamine pratensis* in his *Species Plantarum*.[55] His technique was to classify plants by the number and arrangement of their reproductive organs. The family Brassicaceae (mustard plants), to which the genus *Cardamine* belongs, falls under the now obsolete rubric of *Tetradynamia* (tetradynamous), that is, having six stamens (male reproductive parts) in a particular formation where four are longer than the other two. Many of the species of *Cardamine* listed by Linnaeus were originally classified as species of *Nasturtium*, including *C. graeca*, which was formerly *N. montanum nanum*,[56] and *C. chelidonia*, formerly *N. pyrenaeorum aquaticum latifolium, purpurascente flore* (aquatic and broad-leafed, with a purplish flower).[57] Bauhin had described *N. pratense magno flore* (with a large flower) and *N. pratense, folio rotundiore, flore majore* (with a rounder leaf and a large flower) in his *Pinax*. The latter Linnaeus renamed and reclassified in the protologue as *Cardamine pratensis foliis pinnatis: foliolis radicalibus subrotundis: caulinis lanceolatis* (with pinnate leaves, roundish leaflets and roots, and lanceolate stems). The only other original information he provided was the habitat, stating that it dwells in the wetlands of Europe (*in Europae pascuis aquosis*). In this polynomial, Linnaeus incorporated the diagnostic features of the species into the binomial epithet with the series of descriptors in the ablative case. This was already a convention in other botanical works, albeit on a smaller scale, using shorter phrases. Despite the ambitious nature of his cross-referencing, however, it is not clear from this entry that *Cardamine pratensis* is the reclassification of *Nasturtium agreste*, because Linnaeus did not cite Fuchs. Modern readers can make the connection that Bauhin cited Fuchs and cross-referenced *Nasturtium pratense* with *Nasturtium agreste*. Linnaeus' readership was not expected to make this same connection: that was neither his nor their concern. But for the historian of botany, it is ironic that the most prolific and authoritative list of plant species was still not enough to positively identify the *herba nuda* that most frustrated Brunfels. At least some of Bauhin's and Linnaeus' readers did make the connection and one can find cross-references written in the margins of countless physical copies of both authors' texts.[58]

For examples of Linnaeus' writing style, the *Genera Plantarum* is a better resource than the *Species Plantarum*. When discussing a particular genus, he listed the various plant parts, abbreviating them so that only a practitioner or someone familiar with the field and its specialized terminology would recognize the term. He regularly, for example, abbreviated calyx to CAL. and pistil to PIST. For each part, the plant was also described in a very succinct manner. For the genus *Sisymbrium*, which was linked by Cresson and Tournefort to *Lepidium*,[59] the name given to *Nasturtium agreste* by Galen and Pliny according to Fuchs, Linnaeus provided descriptions of the calyx, corolla, stamens, pistils, perianth, and seeds. The categories through which he proceeded are the ones he had delineated in his *Ratio Operis*: they constitute the flower parts used for plant classification. Where Bauhin adopted a syncopated writing style that was heavy on abbreviations and simplistic grammar and syntax, Linnaeus dropped all pretense and used exactly and only as many words as were necessary to make his point, listing the parts that are used to diagnose the plant genus and describing each on its own line, distinguishing plants in this category from others. The vocabulary is also no longer simply more technical. It is exclusively botanical. The calyx is therefore described as,

> *Perianthium tetraphyllum: foliolis lanceolato-linearibus, patentiusculis, coloratis, deciduis.*[60]
>
> Having a perianth that is *tetraphyllus* and with leaflets that are *lanceolatus-linearis*, *patentiusculus*, *coloratus*, and *deciduus*.

That is, it is four-leafed, with leaflets that are straight and pointed, broad and colored, and which shed in season. Stylistically, Linnaeus did not compare most of the parts (calyx; corolla; stamen; pistillum; pericarpium; semen) to other plants. Rather, he stated the features of each one, using botanical terminology and ablatives of manner and description. There are almost no verbs and those that do appear are almost invariably in the indicative mood. Under the notation OBS (observations), he used complete sentences with neutral word order (subject in first position and verb in last position), either stating further details about the structure of the species in question or explaining how it can be distinguished from others. The formula "species 1 is distinguished from species 2" was copied from Linnaeus to such an extent that by the twentieth century it had become a *de facto* rule for composition of botanical diagnoses. Linnaeus' writing style is not, however, simply an advancement of Bauhin's style. He was able to write in such an efficient manner, using so much notation, because the plants in his books were already known to his audience. At the beginning of *Genera Plantarum*, a section called *Quaerenda* (Queries) is addressed to his readers, whom he assumed to be fellow botanists. Here, he listed a number of plant genera about which he was unsure of the proper classification.[61] Were Linnaeus solely engaged in a project of transmitting his own *cognitio* (knowledge) about plants to his readers, he would have no need of such a dialogue with them. Nor would there be any need for it were he engaged only with the works of prior botanical authorities. By inviting the opinions of his readers, Linnaeus made it clear that plant knowledge is both communal and dynamic. At the same time, the process of classification, naming, and knowing relies on the prior knowledge of the community of botanists. This is why Linnaeus used the adjective *nota* (known) to refer to plants (*vegetabilia*). He was not introducing new species to the world. Rather, he was honing the ways in which plants were already known.

Post-Linnaeus

Linnaeus' influence can be seen in the almost immediate adoption of his style in texts such as *The Botanist's Repository for New and Rare Plants*, published serially in ten volumes from 1797 to 1812 by Henry Cranke Andrews, an English botanist, illustrator, and engraver. Andrews was a skilled artist, not only drawing and engraving the plates himself, but also coloring them in. The *Repository* is also unique because it was sold for a reasonable price, half a crown, which opened the world of botanical knowledge up to amateurs who may have been excluded by institutions such as the Royal Gardens, Kew. His illustrations are accomplished, though they drew criticism by botanists, who saw too much artistry in them and too little accord with the scientific descriptions.[62] However, Andrews had studied his Linnaeus and his textual descriptions are almost direct copies of his style. He classified *Hibbertia volubilis* (Twining Hibbertia) according to its reproductive organs, as *Polyandria Pentagynia.* Immediately, the reader of the *Repository* would have realized how Andrews differed from other botanists, because he provided an English translation of this classification: "Many Chives. Five Pointals."[63] This translation is now outdated; *polyandria* is "many stamens" and *pentagynia* is "five pistils," but this is not important. What matters is that Andrews knew his audience and recognized that curious gardeners and amateur botanists might not know botanical Latin or be familiar with Linnaeus' system of classification. Not only did he translate the terms, but he placed the translation right next to the Latin, allowing his readers not only to understand what the scientific terminology meant, but to learn in the process. As a reader encountered these Latin phrases, their eyes would immediately glance to the English and back again, a solid pedagogical practice.

For the diagnostic features of the genus, Andrews copied Linnaeus again and listed the parts individually in order: calyx, corolla, stamina, pistilla, pericarpium, and semina. These terms he translated as "empalement," "blossom," "chives," "pointals," "seed-vessels," and "seeds." For a different audience, these definitions would not have been necessary, but Andrews ensured that his readers could follow along at every step. The plant parts are described in Latin with ablative constructions and technical botanical vocabulary. For example, the calyx (empalement) has the following information: *Perianthium pentaphyllum; foliolis coriaceis, sub-ovatis, concavis, acutis, persistentibus.* This Andrews translates as "Cup five leaves; leaflets tough, nearly egg-shaped, concave, sharp, and remaining." The only term that is confusing is "cup" for perianthium, now generally referred to as the perianth, the non-reproductive outer structure of a flower, consisting of the calyx (sepals) and corolla (petals). Similarly, a modern translation would simply use "persistent" instead of "remaining." The rest is straightforward, although it is unclear why he also preserved the elliptical, economic precision of the scientific Latin. He may have wanted the amateur botanists who read it to familiarize themselves with this syncopated style. For the species, the diagnosis consists of one line: *Hibbertia foliis obtuse-serratis, obovato-lanceolatis, floribus solitariis, sessilibus, foetidissimus.* The translation provided by Andrews reads, "Hibbertia with obtusely sawed leaves, between lance and inversely egg-shaped; flowers grow solitary, sitting close on the stem; very foetid." In this case, Andrews began by imitating the syncopation of the Latin, but once he reached the flowers, he abandoned the attempt to render the ablatives in English. He resorted to simple indicative statements, but the precision of Latin is lost at a crucial point. The superlative adjective *foetidissimus* is masculine, nominative, singular, and therefore should not be describing the leaves. The only terms to which it could be referring based on gender are the name of the genus itself, *Hibbertia*, after George

Hibbert (1757–1837),[64] and the flowers, since *flos* is masculine. Grammatically in Latin, an adjective modifying a noun will also agree with it in case and number, so if *foetidissimus* is describing the flowers, it should be an ablative plural like *floribus*. English is not inflected to the same degree, however, and it is word order that indicates relationships between terms. The placement of "very foetid" right after the statement about the flowers makes it clear that they have a strong scent, and either the Latin is unclear or the correct form *foetidissimis* is misprinted. Regardless, Andrew's *Repository* is a refreshingly accessible example of botanical Latin that accomplishes the goals of communicating scientific information about plants while acknowledging the limitations of this method.

Another example of post-Linnaean botanical Latin can be found in the 1838 *Elenchus Plantarum Novarum* (List of New Plants) of the Swiss botanist Edmond Boissier (1810–85). In his entry for *Quercus alpestris* (Alpine Oak),[65] Boissier employed a similar structure as Clusius, with a few differences. He provided exactly one name for the plant, which consists solely of the genus and species name. The polynomial format employed by earlier botanists and recommended by Linnaeus did not survive the intervening century. Instead, the ablative constructions have been dissected out into a lengthy diagnosis taking up five entire lines of the page. Almost every word of these five lines is a technical, botanical term and they all hang off of a main clause where the verb is elided. Boissier began with the succinct statement: "A tree 20–30 feet high" (*Arbor 20-30-pedalis*). The remainder of the diagnosis consists of concise statements about the identifying features, beginning with the trunk, which is "rugose and only slightly fungous" (*trunco rugoso minime fungoso*); proceeding to the branches, which are "pubescent"; and on to the leaves, which are afforded the most detail. These are described as "pale green, oblong-lanceolate, deciduous, coriaceous, glabrous above and tomentose underneath, venous; undulate on the margins, and with broad, sharp teeth that are well-spaced but irregular."[66] The cupula, a formal term for a cup-shaped cap surrounding the fruit, common in oaks, has "appressed, tomentose scales," while the nut is "small, ovate, and obtuse with a sharp point." The transition to a more contemporary iteration of botanical Latin is clear in this entry. The string of ablative descriptors modifies the noun *arbor* with a large number of adjectives, most of which are fairly technical, and all of which follow a strict pattern of "with x feature." Each of these adjectival descriptors should be taken as essential to the diagnosis of the species. To the non-practitioner, however, only a handful of words in the entry will make sense: *arbor* (tree); *foliis* (leaves); *deciduis* (deciduous); *irregulariter* (irregularly); and perhaps *dentibus* (teeth), since the term *dentatus* (or *dentata*) just means that the margins of a leaf have sharp edges that resemble teeth. The other terms will need explanation to a non-practitioner, and perhaps even to someone who is familiar with classical or medieval Latin, but unfamiliar with botanical Latin. *Rugose*, for example, means "wrinkled; having ridges," while *tomentose* means "woolly; covered with a fine mesh of hairs," and *fungous* means "spongy." While it could be argued that a trained classicist could surmise these definitions, such a stance misses the point. The closer botanical Latin gets to the twentieth century and to a professional field populated by individuals from non-European backgrounds with little or no linguistic connection to classical Latin, the more exclusive this language becomes. Right when botanical Latin reaches its economic and efficient apex in terms of scientific expression, its intelligibility begins to wane. One simply cannot expect anyone untrained in the profession to understand what is being stated and presented as knowledge about this species of oak.

Following this onslaught of technical terms, the second half of Boissier's diagnosis continues with the ablative constructions, but is focused more on comparison of the leaves with other species of oak. He contrasts the leaf texture with that of *Q. pseudosubere* Desf., which is similar to *Q. Alpestris*, but it has fungous (spongy) bark and its leaves are ovate or oblong, not sharply *dentate* (toothed), but with even margins and obtuse and regularly spaced "teeth." Boissier also distinguished *Q. Alpestris* from *Q. faginea* Lam., which is somewhat confusing because the latter nomenclature is now the preferred synonym of the same species. Boissier claimed that the former's leaves have similar curly margins to the latter, but that they are smaller and more rounded (obovate), shiny above, with regularly and very acutely serrate teeth. The leaves of *Q. faginea* Lam. are smaller than those of *Q. Alpestris* and closer together, with secondary nerves underneath that have little projections that are more numerous and straighter, with typical venation. He also stated that the nut of *Q. faginea* Lam. is longer and pointed. It is clear from this level of detail that Boissier did inspect a living specimen of *Q. Alpestris*, but it is unclear how he surmised that it was not the same as *Q. faginea* Lam., unless he was also mistaken about the identification and in particular, about the leaf structure and form of that species. It is possible he was actually observing another species entirely. There is a hint in the last sentence of the diagnosis, "I did not see flowers" (*Flores non vidi*). This indicates that the specimens he saw were either not yet in bloom, or senescing. Next, Boissier transitioned to the extended description. In contrast to the diagnosis, he was now free to provide non-defining, non-diagnostic features of the organism, some or all of which it might share with others. As would soon become standard, the description has a straightforward, narrative feel and is considerably more legible, with unelided finite verbs in the indicative. He frankly noted that he had only observed the tree on the mountain named *Sierra de la Nieve* where it grows in very great abundance with *Abies pinsapo* and maintains an upper limit much higher than any other Granadan species of this genus, at an altitude of three to six thousand feet.[67] Boissier switched to a more modern prose form in which various characteristics are stated in full sentences, each with a noun and verb, but using an ablative of means. In the diagnosis we are given the characteristics with which to recognize the organism, whereas in the description we are given the characteristics by means of which we can recognize the species. This is not a mere stylistic difference. It serves the purpose of letting the reader know that the former characteristics are precisely those which define the organism, and with which it is classified in the plant kingdom. The latter are simply used to further describe it. This distinction between the uses of the ablative was perfectly clear to Boissier, but not so to many twentieth-century botanists, who, in attempting to mimic his style, inadvertently composed their descriptions entirely in the ablative, exactly as one would the diagnosis, thus missing the point of using the ablative of means in conjunction with the nominative forms of the plant's organs.[68]

Conclusion

These passages are not meant to be fully representative of the state of botanical Latin at their particular time periods. Rather, by looking at them as textual objects of value in and of themselves, it is possible to collect data that can be contrasted with the proto-botanical Latin of the sixteenth century. There is considerably more technical vocabulary in Boissier, where a large percentage of the vocabulary has specific botanical definitions, than in earlier texts, even Bauhin and Linnaeus. Yet the new botanical terms themselves are often simply appropriated from either classical or medieval Latin, so that the botanical definition can sometimes be discerned in the words, assuming one has prior knowledge of Latin or Greek. The divide between twentieth-century botanical Latin and the earlier iterations is most significant with Clusius, whose entry just barely made it to the seventeenth century and whose use of specialized technical vocabulary is minimal compared to the later entries. It would be neither possible nor methodologically sound to draw a single, causal line between Brunfels and the kind of botanical Latin seen in Clusius or Linnaeus, any more than it is possible to set a specific date for the advent of botany as a science. The sixteenth-century botanical texts also employ "botanical Latin," even if, unlike these later examples, it is difficult to separate out from other genres of Latin stylistics. As a result, these early texts are best seen as part of the transition from the medieval Latin of the herbal tradition to the scientific Latin of the Renaissance. Modern botanical Latin involves the application of pre-established rules and recommendations for the sole purpose of scientifically describing a species or genus so as to distinguish it from others as a distinct entity. In conjunction with the waning of Latin as the universal language of science, this has resulted in a style of composition and a use of technical vocabulary that is distinct from earlier examples of botanical descriptions written in Latin. The progression from Clusius's 1601 description of *Siliqua silvestris*, which not only is clear and precise but demonstrates mastery of Latin as a language of expression, to twentieth-century botanical diagnoses, which are long on technical vocabulary and incredibly short on skill or style, is very little studied. However, in this time frame and with these passages, it is possible to identify the introduction of polynomials constructed with ablatives, their formalization in Linnaeus, and their ultimate rejection in favor of a binomial and diagnostic combination. In addition, we can observe the development of a technical, strictly botanical vocabulary characterized by diminutives matching smaller identifiable plants parts. The writing style becomes syncopated, with verbs eliding in favor of diagnostic and descriptive terms and phrases, and from the cacophony of nomenclatural debates arises the almost palpable relief of accepting Linnaeus' classification system. For practitioners, this was a period of stabilization. By streamlining all of these naming and classification systems, by establishing a specific vocabulary, and by narrowing the ways of communicating scientific information down to a few, standardized formats in a seemingly fixed and secure language, they believed that they were also stabilizing their ways of knowing, their axioms, their methods of discovery, and their adherence to the ideal of universal intelligibility. Andrews saw the potential problems almost immediately, although we must be careful not to assign the same ethical and philosophical concerns to him. He wished to communicate with a specific, English-speaking audience. However, at the same time that botanical Latin was coalescing into its restrictive twentieth-century iteration, there were other botany-related events occurring outside of the confines of Western Europe, which influenced how botanists studied and conceived of knowledge about plants and at the same time, introduced challenges to their long-established patterns and beliefs that caused the botanical community to close ranks and instigated the late-nineteenth-century turn to rules and regulations.

4

Botanical Latin and Nomenclature in the Twentieth Century

Beginning in the late nineteenth century, the international community of botanists began to coalesce into a professional organization with annual meetings at which practitioners could discuss, debate, and vote on issues affecting the field. Despite their self-assigned nomenclature, however, the International Botanical Congress was primarily Western European and North American in its makeup. A complete understanding of how botanical Latin transformed at the time into a rigid set of rules for naming and describing plants depends not only on recognizing that the science of botany excluded non-Western practitioners, but also that the exploration and colonization of the Americas affected the development of botany and botanical Latin.[1] Botanical expeditions were about control and power and it is no coincidence that as Europe was losing its grip on its many colonies, botanists were convening to form the first professional rules and regulations for expressing knowledge of plants. European botanists had already attempted to apply their newly honed methods to the study of plants in the so-called "New World," where much of the flora was unknown to them. Despite this initial "rhetoric of wonder," plant knowledge, especially medicinal, was quickly commodified and commercialized. In the process, European botanists, and those who funded their expeditions, immortalized colonizers in the nomenclature, rejecting indigenous names entirely or Latinizing them to fit in with binomial nomenclature. They were also, at least initially, so awed by the strangeness of these new plants that they retreated into their increasingly formal scientific methods: the chaos and instability of life in the Americas was assuaged and lessened by subjecting the flora and fauna to classification and analysis. The need to describe plants in the language and style of Europe, for Europeans to understand, however, put to the test the extent to which botanical Latin and nomenclature promoted discoverability, knowability, and intelligibility, ideals that were frequently ignored as they pertained to indigenous people. The development of botany as a scientific discipline on the continent is more fully understood when it is contrasted with botanical expeditions that were funded by imperialists with entirely different goals than locally trained botanists restricted to the study of plants in their own locales. Where European botany grew organically out of the herbal tradition, which was focused almost exclusively on the medicinal properties of plants, in the Americas, even when an explorer or botanist took the time to describe the morphological characteristics of a plant in addition to its medicinal properties, that information was almost invariably proprietary, the "property" of whomever funded the expedition, usually a political entity. Imperialism contaminated any purity that might have characterized the study of nature. At the risk of oversimplification, descriptions of plants in the Americas were fundamentally either commercial or colonialist, exploitative and dismissive of local populations, and overtly exclusive. In keeping with transformation methodology, the development of botanical Latin can therefore be understood as an *allelopoietic* reception sphere reacting to botanical colonization, and vice versa.

Plant Exploration in the Americas

The increasing formalization of the methods of describing, discovering, and transmitting information about plants that began in roughly the sixteenth century had produced its own subspecies, which, from a methodological standpoint, can be viewed as another cluster of points on the continuum of the development of botany and botanical Latin. Taken together, these points comprise the botanical colonization of the Americas and other countries, especially those in South Asia and the African continent. In the former case, the term "New World" persists, perpetuating the idea that the natural world in the Americas was unknown and undiscovered prior to the arrival of Europeans. Few would argue that plants indigenous to the Americas were literally previously unknown, but in most European expeditions, the stance was clear: *naturalia*, plants included, were not "known" in ways approved of by Europeans. Linnaeus and other European botanists influenced the observation, description, and naming of plants in the Americas and other colonized regions, but at the same time, their own processes were affected by the realities of studying nature abroad.

Many of the early expeditions in the fifteenth century involved the search for trade routes in addition to new land to occupy.[2] England alone was responsible for the infamous Roanoke Colony established in 1585 by Sir Walter Raleigh (*c.* 1553–1618); the Jamestown settlement of 1607, which developed into Virginia;

Massachusetts, settled by Pilgrims in 1620; Bermuda, which developed from a shipwreck in 1609; and, throughout the seventeenth and eighteenth centuries, various colonies in the Caribbean such as Jamaica and Barbados, and the areas surrounding the Hudson Bay, where New Amsterdam, later New York, arose. The Dutch, although they had settled some areas of mainland North America in the seventeenth century, including the Chesapeake and Narragansett Bays, did not maintain their stronghold. The Dutch Republic (*Republiek der Zeven Verenigde Nederlanden*) had in 1602 chartered the Dutch East India Company (*Vereenigde Oostindische Compagnie* or "VOC") to find the much-sought Northwest Passage to the Indies. Once the English, French, and Spanish took control of most of the mainland,[3] the Dutch turned their attention to Asia, specifically Indonesia, where the VOC established trading posts. Meanwhile, France established colonies all over North America, the Caribbean, and South America, but their first expedition was in 1524, when King Francis I sent Giovanni da Verrazzano (1491–1528) to explore almost the entirety of the East Coast in pursuit of a passage to the Pacific Ocean. The French maintained control of several areas in the Hudson Bay and Hudson River Valley, including parts of Brooklyn and Queens, before the English and Dutch took control. The first overt attempts at colonization and the establishment of permanent settlements began in 1534 when Francis I sent Jacques Cartier (1491–1557) on three different voyages to eastern North America, mostly modern-day Newfoundland and Quebec. Spain in the meantime had established Hispaniola in 1493 on the island now known as Haiti and the Dominican Republic. A year later in 1494 they signed the Treaty of Tordesillas, which divided the "New World" between Portugal and the Crown of Castile.

In exactly the same time frame in which botanical Latin was formalized in mainland Europe, a process made possible in part because the plants being discussed and named were already known and discovered, there was a steady flow of information from all over the world about "new" and "unknown" plants reaching these same botanists and herbalists. By the sixteenth century, the reasons for at least some of the European expeditions had changed from the pursuit of trade routes or attempts at permanent settlements to the natural historical. Some were overtly botanical, focused on collecting, observing, describing, and illustrating medicinal herbs and any other flora with potential utility and trade value. Initially at least, the sheer newness and vastness of all this *naturalia* almost seems to have shocked the Old-World naturalists into old habits of expression. In a 1670 letter, the English Puritan John Winthrop described how he had winced when confronted with the strangeness of nature in

Massachusetts.[4] In the early American naturalist tradition, which grew out of the collision between European intellectual history and the same landscape that caused Winthrop to blanch, Linnaeus' injunction that the study of nature should be "simple, beautiful, and instructive,"[5] provided the impulse to begin compulsively collecting nature's products. The unease resulting from interaction with nature's strangeness was somewhat alleviated by careful inspection and by application of Linnaeus' methods, particularly his standards for binomial epithets. Yet Linnaeus was a century or two away for the early explorers and colonizers and in the meantime, they had to find methods of coping. Knowledge of how to survive in this new landscape, knowledge that the indigenous people already had, was deemed inferior because it was not obtained in scientific ways. Winthrop's unease was unnecessary and betrays his distrust of indigenous knowledge. Overt racism and a paternalistic view of indigenous people played a significant role in the stubborn insistence of the Europeans on long-distance communication and their retreat into familiar ways of knowing. Recognizing this also helps to explain the "rhetoric of wonder" assigned to New World naturalists. Rhetoric is not concerned with objective truth, but convincing others of the truth or validity of one's argument. The awe and wonder expressed by Europeans in the Americas was generally not about appreciation; it was about not initially knowing how to capitalize on or appropriate this unfamiliar Nature. The skepticism of indigenous knowledge can be seen in John Gerard's 1597 *The Herball*, which includes a number of American plants such as the potato and what he calls alternately *Frumentum Indicum* and "Turkie corne." Gerard claimed that corn is a kind of wheat but that it is less nourishing than regular wheat, rye, barley, or oats, citing the fact that the Native Americans, who do not know any better,

> are constrained to make a vertue of necessitie, and think it a good food; whereas we may easily judge that it nourisheth but little, and is of hard and evill digestion, more convenient food for swine than men.[6]

Gerard consistently referred to various tribes derisively, grouping them all under the term "Indians." He operated from the implicit assumption that anything they did was wrong and that he and his fellow Europeans knew better. Prior to this, the French explorer Jacques Cartier (1491–1597) had made three voyages to modern-day Canada in search of the Northwest Passage, the first in 1534 to Newfoundland and St. Lawrence and surrounding regions, where he planted a flag claiming the land for the French throne and took captive two sons of the Iroquois chief. He brought them to France under the condition that he would return with European goods, which he did in 1535–6. Over that winter, when his fleet was iced-in near the mouth of the St. Charles River, an outbreak of scurvy resulted in the deaths of dozens of Iroquois and French sailors. It was from an Iroquois man that he learned of a cure, a tea made from the bark of a tree called *annedda*, without which his crew would not have lasted the winter.[7] It is unclear exactly which tree this was. Some identify it as *arbor vitae* (*Thuja occidentalis*) or white cedar, while others posit spruce (*Picea*) or hemlock.[8] In 1781, Jonathan Carver, in his account of his travels in North America, described the Sweet gum tree, also called Liquid Amber (*Copalm*), which "affords a balm, the virtues of which are infinite" and which is "reckoned by the Indians to be an excellent febrifuge, and it cures wounds in one or two days."[9] Carter rarely bothered to mention the name of the tribe about which he was speaking and limited his physical descriptions to the absolute basics, such as a general leaf shape and the quality of the wood.

From these early first-hand accounts of American flora, it is clear that there was no concern for taxonomy or classification. Neither, however, was there any serious concern for identification in the wild. It became common to obtain and send back to Europe specimens of flora and fauna, many of which dried out or died on the passage over, prompting developments in the transportation of live specimens and, for plants, the preservation of seeds for propagation. John Bartram (1699–1777), one of the earliest American botanists, sent plant specimens and seeds in "Bartram's Boxes" every autumn to various European botanists via the English merchant Peter Collinson. It is through Bartram that European botanists learned of such flowering plants as the rhododendron and magnolia. The latter genus and its known species are described by Linnaeus in his *Species Plantarum* with the summary *habitat in Virginia, Carolina* (It grows in Virginia and Carolina).[10] The problem of transporting live specimens was not solved until the nineteenth century, when the London-based botanist and physician Nathaniel Bagshaw Ward (1791–1868) developed a kind of vivarium or mini-greenhouse to protect the plants in his collection, especially his ferns, from pollution.[11] The utility of these "Wardian" cases for long-distance transportation was quickly realized and plant specimens were soon able to be preserved in living form, allowing for more accurate observation of their morphology and growth patterns.

Despite the application of Linnaeus' system of classification and recommendations about nomenclature to the flora of the Americas, botanical Latin did not really exist outside the confines of the European continent until the late nineteenth century. This may seem unlikely, since one of the primary characteristics of the colonial era was collecting *naturalia* worldwide, botanical specimens in particular. However, the inability to preserve specimens and the difficulties of propagating from seeds meant that botanists were unable to diagnose plants, to recognize and describe their defining characteristics. Without a diagnosis, they could not accurately classify plants, identify their taxonomic rank, or determine if they were entirely novel or a different species of a genus already endemic in parts of Europe. This imprecision and uncertainty did not preclude the use of botanical Latin, but given that it was already developing into a highly technical vocabulary strung together with idiosyncratic syntactical constructions, it was simply of little use for the description of the thousands of new plants being collected. Leonhardt (2013) has pointed out that many educated Europeans had a certain mastery of Latin much as the elite in ancient Rome were familiar with Greek.[12] Yet Latin was reserved for formal writing and most botanical texts produced in these circumstances were written in the vernacular, in Spanish, Italian, French, and English. While Latin had become the language of formal scientific description in Europe, observations in the field abroad were often done in one's spoken language. There are exceptions, but writing in Latin is not the same as using botanical Latin; one is a natural language and the other is a scientific tool that for a long time was unwieldy in the field. The task of converting *ad hoc* notes, remarks, and illustrations into publishable form was either left to the scientists in Europe to whom specimens were sent, or it was handled by explorers themselves at a later date, when perhaps they had more leisure to correspond with one another and consult the printed botanical texts that served as their references and sources of authority. It was in part this reliance on the vernacular that prompted a turn away from textual diagnoses and descriptions and toward visual forms of communication. Sketches, illustrations, paintings, and specimens, both living and dried, became far more useful and necessary means of communication than botanical Latin.

Where we do see Latin being used, it is almost invariably due to the influence and training of the Jesuits or other Holy Orders, whose frames of reference tended to be either religious or classical in nature. In other words, while the Jesuits were teaching Latin to indigenous people, it was not scientific Latin.[13] The Franciscan friar Jerónimo de Aguilar, who was shipwrecked in 1511 on an island off the coast of the Yucatán and enslaved by the Mayans until the conquistador Hernán Cortés arrived in 1519, is often credited with accidentally introducing the Latin language to México. He apparently still had his Book of Hours in 1519[14] and it is assumed that he daily recited his prayers while in captivity, thus exposing the Mayan people to the language. Latin was the language of the Church, law, science, literature, and of classical antiquity and it was inevitable that with the conquistadors came their languages, the vernacular and formal Latin both. Part of the colonization process involved religious conversion, which was effected by the establishment of San José de los Naturales in 1527 and Santa Cruz de Tlatelolco in 1536, two colleges that trained indigenous Mexicans in liturgy and European intellectual pursuits, in Latin and Nahuatl. Botanical Latin, however, was in its infancy and not a high priority for the colonizers. Despite this, when they began naming and describing nature's productions, European and American explorers, naturalists, and botanists still contributed to the development of formal botanical Latin by Latinizing the vocabulary[15] and nomenclature of the vernacular and of indigenous languages. The primary effects of the formal botanical Latin that was still developing in continental Europe therefore lay in the encoding of colonizers in nomenclature and in the Latinization of indigenous names. In his *Rusticatio Mexicana*, an epic poem in meter composed in Latin, Rafael Landívar (1731–93) provided lengthy descriptions of the flora and fauna of South America, but these were literary passages, not diagnostic. It is the Latinized names and terminology that stand out, for example in Book IV, titled *Coccum et Purpura*, in which he described the cochineal insects (*Dactylopius coccus*) from which crimson lake (carmine) dye was derived. Landívar called the host plant *Nopalis*, a Latinization of *nopal*, from *nohpalli*, a Nahuatl word for *Opuntia* (prickly pear) cacti.[16] Landívar's approach was also distinct from that of the explorers. He was from Guatemala and had seen the dismissive attitude of Europeans toward the Americas and its people. Laird (2006) posits that Landívar may have been prompted to compose the *Rusticatio Mexicana* by the publication in 1768 of Dutch cleric Cornelius de Pauw's *Recherches Philosophiques sur les Américains*, in which he argued that there was no book produced in America worth reading, that indigenous Americans were degenerate and child-like, and that any evidence of their intelligence, such as Aztec calendars, must be forgeries made by European explorers.[17]

One of Landívar's natural historical influences was Francisco Hernández de Toledo (*c.* 1515–87), court physician to King Philip II of Spain and a trained naturalist, who led a years-long expedition to the "New World," from 1570 to 1577, having been appointed chief physician (Archiater) to those regions. Hernández spent the first three years travelling around México and Central America before settling in México to study the herbal medicinal knowledge of the Nahua people, working closely with indigenous artists, healers, and others with extensive first-hand knowledge of local herbs. He is said to have been inspired to include copious illustrations in his natural history by the "philosopher king" Nezahualcoyotl of Texcoco, who had his own visual inventories of local flora and fauna.[18] Hernández apparently spent this time not only collecting specimens and illustrations,[19] but testing out the medicinal properties of the herbs. In addition to herbs with curative properties, Hernández studied toxic and hallucinogenic flora such as the *Strychnos nux-vomica* tree (Nahuatl: *Mahuatl Quauhtlepatli*), the fruit of which produces strychnine, and peyote (from the Nahuatl *peyōtl*, "caterpillar cocoon"). He also studied the *Datura* genus of the Nightshade family Solanaceae, species of which, often called Moonflowers or Devil's Weed, were used as psychotropics. Hernández is also credited with first describing pineapples, corn, cocoa (*Cacahuatl*), passion fruit, and others for a European audience. When he returned to Spain in 1577, he had gathered some thirty-six volumes in manuscript form, of which ten were illustrations, two dealt with minerals and fauna, and the rest focused solely on the flora of the region. He died before these could be printed, having only produced an introduction. After his death, the Archiater of Napoles, Nardo Antonio Recchi (d. 1595), was commissioned by Philip II to compile the manuscripts for publication, but he too passed away before he could do more than arrange the plants by morphological similarities and their possible utility for diseases that were plaguing continental Europe. A Dominican priest by the name of Francisco Ximénez (*c*. 1560–1620) obtained a copy of Recchi's work[20] and it was finally printed in México in 1615 under the title *Quatro libros de la Naturaleza*.[21] This far pre-dated Linnaeus and his authoritative classification of the plants known in Europe and there is no evidence of the European impulse to immediately assign Latin nomenclature to indigenous names. Recchi had organized the plants according to their morphological features, since that was essentially all he had to work with; most of the plants Hernández described were unfamiliar to Europeans. The first category in the text is *Coapatli*, also called *tlamacazquiypà*. Herbs, he notes, are called *coapatli* if they have leaves that are similar to those of a peach (*ojas semejãtes a las del durazno*), produce curly, rounded stems (*tallos redondos y rolliços*), and have small, cupped, white flowers that turn yellow at the tips (*las flores acopadas y pequeñas, y en los estremos blancas tirantes amarillas*).[22] Not all of the herbs are entirely unknown. The herb called *Pehuame* by those in Mechoacan has the same medicinal benefits for childbirth as the plant Dioscorides called *Aristolochia*. It is described as having small purple flowers (*flores purpureas*) that are quite similar to other kinds of *Aristolochia*.[23] The cocoa tree (*Cacahuacintli*) is described by reference to an orange tree, being similar in size and leaf shape, though cocoa leaves are much larger and more numerous. Its fruit meanwhile is likened to a large pepper, albeit wrinkly and red in color (*acanalado y rojo*).[24] Much of Hernández' original manuscript was lost in a fire in 1671, but we know, despite some contradictory accounts, that he originally wrote in Latin. The Spanish-language printed edition that became his legacy, then, was not the original intention. Given his preservation of Nahuatl nomenclature, it is likely that he planned to have his text printed in Latin, perhaps with Nahuatl translations, much like the Cruz–Badiano manuscript of 1552. Also called the *Libellus de medicinalibus indorum herbis*, the Cruz–Badiano manuscript contains Latin translations of Nahuatl descriptions of Mexican medicinal herbs. The Latin translations were made by Juan Badiano, with the Nahuatl text provided by Martín de la Cruz. The manuscript retains the Nahuatl names, providing Latin names only when the plant is thought to be the same as one known in Europe, making no attempt to Latinize them. To a degree, this is simply because botanical nomenclature was still developing and the set of formal, agreed-upon names, was still quite small at this point. Yet Badiano also took this approach with the Latin descriptions and any time a Nahuatl word could not be translated, he simply transcribed it into Roman letters. This conservative approach to indigenous knowledge, of not disdainfully rejecting it wholesale, was one of the criticisms leveled against Hernández, who by some accounts displeased the Crown by deferring too much to the indigenous population. Benito Arias Montano (1527–98), editor of the *Antwerp Polyglot* and Royal Chaplain to Philip II, reportedly complained that Hernández had been too integrated with the Mexicans and had failed to convert them to European ways.[25] It is quite likely, then, that his original plan had been to publish in both Latin and Nahuatl.

During the time Landívar was working, the Spanish Crown funded some eight natural history expeditions in the eighteenth century, a mere fraction of the scientific campaigns it sponsored in the name of political and economic expansion and augmentation. In direct competition with other European nations and in the hopes of revitalizing the country's economic situation, Spain turned its attention to more lucrative commodities in the Americas, such as coffee, spices, tea, pepper, nutmeg, and so-called *materia medica* (natural medicinal remedies).[26] Prior to this, Spain had been interested in minerals, but it had realized it could grow these commodities in their own colonies, reducing their need to rely on other sovereign nations and allowing them to set up near monopolies on certain markets. The Sevillan doctor Nicolás Monardes (1493–1588) had, for example, been the first to produce a herbal of American plants in three parts, the first appearing in 1569, in which he discussed *Mechoacan*, a plant with strong medicinal benefits from New Spain, from which it received its name.[27] Monardes also was convinced of the curative powers of tobacco, which he discussed in the second part of his herbal in 1571.[28] The medicinal benefits, real or imagined, of such plants quickly increased in value, and the Spanish Crown eventually saw an opportunity to monopolize their trade and export. The Spanish Crown, unlike other European nations, where such practices were *de rigueur* but informal, made it imperial policy[29] for institutions such as the Royal Botanic Garden in Madrid and the Natural History Cabinet to not only request specimens, illustrations, and descriptions of flora and fauna through networks of naturalists, tradespeople, and explorers, but to turn this into a bureaucratic effort, specifying desired genera and species, numbers of specimens, and even how best to transport them.[30] As Bleichmar (2004) notes, however, despite the demands of the Crown, for those Spanish colonists ensconced in México or South America, distance from the continent resulted in a certain autonomy and a shifting of priorities and allegiances. Spanish colonists began to understand the natural world around them in different ways, in particular, by visual investigation, observation, and communication.[31] As a result, while specimens and textual descriptions of plants were still valuable, illustrations became a crucial part of their identification and appropriation. One of the few European women naturalists, the German entomologist and artist Maria Sibylla Merian, traveled with her daughter Dorothea to Suriname, then under Dutch rule, in 1699 to study and illustrate the insect populations, producing in 1705 *Metamorphosis Insectorum Surinamensium*. The text, written in Dutch, is remarkable for its full-color, full-page illustrations of insects and plants and for the quality of the textual descriptions, including her observations on the metamorphosis of butterflies. Merian's work is also conspicuous for her considerably more empathetic view of the people of Suriname, especially the women, who used the peacock flower (*Caesalpinia pulcherrima*) as an abortifacient, providing the following observation with the illustration:

> The Indians, who are not treated well when they are in service with the Dutch, prevent their children from being born into it, not wanting their children to be slaves. The Black slave women from Guinea and Angola must be treated very kindly, else they desire no children in their enslaved state and in fact have none.[32]

Illustrations such as Merian's not only furnished a European audience with invaluable information, they traveled better than specimens and were less prone to misunderstanding than textual descriptions. Standards for illustrations also developed, as did best practices for observation, specimen collection and preservation, and description. As naturalists on the continent began to produce printed texts about these plants, those embarking on expeditions began carrying these new books with them for reference, thus closing a circle they were unaware of. Where initially they might have buttressed their descriptions and illustrations with knowledge gained from local populations, eventually, they had produced their own body of knowledge in their own languages and they treated that information as the foundation for their continued experience with the plants *in situ*.

An example of this approach can be seen in the quarter-century long botanical expedition conducted by José Celestino Mútis (1732–1808), a Spanish priest and polymath. Mútis had begun petitioning the king to fund an expedition in New Granada (modern-day Colombia, Ecuador, Panama, and Venezuela) beginning in 1783, after he had already been there serving as the physician to the viceroy since 1761. Over the course of twenty-five years, while in correspondence with Linnaeus, he systematically collected specimens in the field and compiled data on their habitats, growth patterns, and uses, whether medicinal or otherwise. He also accrued some 8,000 plates, including maps and illustrations made by his pupils, 24,000 dried specimens, and a variety of other *materia*, including bark, wood, resin, shells, and more, much of which was sent to Spain, to the Madrid Botanical Garden. Once there, they languished uncatalogued, unanalyzed, and uncollated, highlighting the reality of this methodology, which is that careful collection and description have no use if they are not followed through on. Despite this, his work on a particular genus, *Cinchona*, is well-documented and we have multiple watercolors sent by him to Linnaeus depicting various species. In a watercolor dated September 1764, before the official expedition funded by the king, the species *Cinchona Peruviana* is drawn *mutis in literis* (with the trunk depicted) in a couple of different stages of flowering (see Figure 4.1). The quality is not comparable to that of the illustrations produced by artists and botanists on the continent, but there is sufficient detail to aid in identification and to allow for cross-referencing with dried specimens and textual descriptions. It can also be compared with a 1742 drawing of *Quina Quina Prima*, the Quechua name for *Cinchona* from which the term "quinine" derives (see Figure 4.2).[33] In this illustration, rather than the entire plant, or even the flowers, one single branch is depicted, with two cross-sections, which appears desiccated and rough and hardly resembles any known species of the tree. The artist J. Hawkins drew this from specimens sent to the Secretary of the Royal Society, which explains the gap between this image and others drawn from living specimens.[34] The name *Cinchona* itself was provided by Linnaeus in 1742. The story, first recounted by the Genoese physician Sebastiano Bado, is that the Count of Chinchón, Luis

Jerónimo Fernández de Cabrera Bobadilla Cerda y Mendoza, was viceroy in Perú when his wife, Ana de Osorio (1599–1625), fell ill in 1638 with malaria and was treated successfully with *quina quina*. According to Bado, she brought the plant back to Europe the following year. This was the account with which Linnaeus was familiar and he used the Italian spelling of Chinchón to name the genus.[35] Linnaeus classified it as *Pentandria Monogynia* (having five stamens and one pistil), and described the calyx, corona, stamens, pistil, perianth, and seeds in the same succinct manner as other genera. Under the heading *Stam.* (Stamens), he states, "*Filamenta quinque, minima. Antherae oblongae, intra tubum corollae*" (Five filaments, small. Oblong anthers, inside the tube of the corolla). Despite the spare nature of this diagnosis, it can be compared to the illustrations of the time period, especially the one sent to him by Mútis. It is also quite telling that this example of botanical Latin was not composed in Perú but in Europe by Linnaeus, after he had had time to correspond with Mútis and inspect specimen and illustration both.

The medicinal benefits of *Cinchona* are wide-ranging; the bark produces quinine (hence the inclusion of a cross-section of the trunk in Mútis' illustration)

Figure 4.1 *Cinchona Peruviana*. Reproduction of a watercolor illustration sent to Linnaeus in September 1764 by José Celestino Mútis. Credit: M0001350. Courtesy: Wellcome Collection.

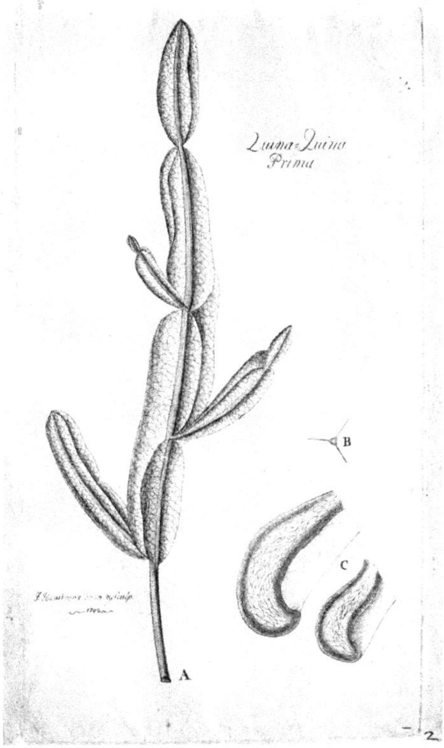

Figure 4.2 *Quina Quina Prima*. Drawn from a specimen by J. Hawkins in 1742. Credit: M0001394. Courtesy: Wellcome Collection.

and other alkaloids that were once the only effective antimalarial treatments. The Jesuit missionaries in Perú, Bolivia, and Ecuador learned of its medicinal uses from the Quechua, who also utilized the bark in sweetened "tonic" water for diarrhea and shivering, associated with malarial fever. Since malaria was not unique to South America and was also endemic to areas of Europe, it is not hard to see why the Spanish Crown was interested in the cultivation and commercialization of such a plant. Species of *Cinchona* are endemic in other parts of the world, which meant that the development of quinine[36] to treat malaria had the effect of strengthening the European colonialist grip in those regions, including India and Java.[37] Medicinal knowledge is power, and as Brockway (2002) has argued, global colonization was incentivized by the search for indigenous botanical knowledge.[38] Erichsen-Brown (1979) has posited that the plant-based medicaments and new food sources exported to Europe and Asia from the Americas caused a population explosion that actually drove further colonization efforts.[39] The exact mechanisms by which this was achieved vary, but Cohn (1996) explains how in India, the British were able to control the population by the most banal of methods: extreme bureaucracy in the form of excessive documentation and information gathering via commission reports, censuses, and statistics on "trade, health, demography, crime, education, transportation, agriculture, and industry," all of which were so difficult to interpret that only the British government itself had the skills to do so.[40] Increased efficiency in documentation and methods of communication were precisely the goals of formal botanical Latin. If we consider its development concurrently with increased colonization, we can begin to see the ideal of universal intelligibility as more than just a myth. It was an exercise in exclusivity from the very beginning. Early explorers may have preserved indigenous names, but these were quickly discarded and they themselves were immortalized in scientific nomenclature. Hernández' name is encoded in *Corytophanes hernandesii* and *Phrynosoma hernandesi,* two species of lizard endemic to North and Central America that resemble those described by him. He is also immortalized in the genus of flowering plants *Hernandia*, first dedicated to him by Charles Plumier in 1703 and accepted by Linnaeus into the family *Hernandiaceae* in 1753.

Hernández' motives are well-known, but we cannot say what he wished for his own legacy. For other naturalists and botanists who succeeded him, however, the lure of fame and respect associated with discovery was certainly a driving impulse. This was especially true after Linnaeus re-classified thousands of genera and species and stabilized their nomenclature in the eighteenth century. European botanists no longer had the need for or privilege of naming and categorizing "new" plants. In order to make names for themselves and in order to further the acquisition of scientific knowledge of the plant kingdom, botanists had to look further afield, to describe and categorize plants that were already known to their local communities. In doing so, they also gave these plants new, non-indigenous, Latin names. By discussing a small number of plants, explorers, and botanists with links to the Americas, the gap between botany in Europe and the early encounters with and descriptions of flora in the Americas come into focus. A further distinction lies in the reasons for illustrations, which differed in the two contexts. In Europe, where botany was developing into a scientific discipline, Brunfels' herbal functioned to demonstrate the utility and efficiency of realistic illustrations. In the Americas, due to the physical distances involved and the difficulty of sending live specimens abroad, illustrations became as valuable for communication as textual descriptions, regardless of language. Despite their limitations, illustrations were more universal than text. The use of

Latin as a "universally intelligible" tool for communication became an afterthought, with most textual accounts of American flora and fauna being written in the vernacular and translated into Latin, either *in situ* like the Cruz–Badiano manuscript or *post hoc*, like the abridged Latin edition of Hernández' natural history, which was printed in 1628 under the title *Rerum Medicarum Novae Hispaniae Thesaurus* (A Treasury of Medicaments from New Spain). The use of the vernacular was not unique to the Americas. Despite the widespread use of Latin on the continent since Brunfels in the sixteenth century, some writers persisted in writing in their primary spoken language. For example, in 1846 the botanist Karl Koch (1809–79) published in German an account of his travels in the Caucasus Mountains, Asia Minor, the Caspian Sea, and Armenia.[41] One possible explanation for this persistent use of the vernacular is precisely its own kind of exclusivity, which functions in a different way than scientific Latin. By using the vernacular, European explorers were not simply writing in their first language, but were inadvertently encoding and preserving in their texts a sense of national identity, which lent itself to the end goal of colonization. If the Spanish Crown wished to take control over certain territories, it served their interests to use Spanish for every form of communication; to use Latin was to acknowledge that the information being conveyed was not their sole intellectual property. In cases where an individual did use Latin, as in the Cruz–Badiano manuscript, this decision can perhaps be read as subversive, a way of signaling that the information being communicated was not the property of one nation. Ultimately, the conquest of the Americas had a significant impact on botany, but little effect on the development of formal botanical Latin, except by providing indigenous names to be Latinized. This was the case at least until the late eighteenth century, when Linnaeus was in correspondence with botanists working in various regions outside of Europe. But botanical Latin was first and foremost a European, continental project.

The Institutionalization of Botanical Latin

Apart from nomenclature, the primary effect on botanical Latin of centuries of exploration and natural historical expeditions was the realization for botanists, certainly by Linnaeus' time, that their means and methods of communication were flawed and insufficient. Both practitioners and institutions were gathering massive collections of specimens, diaries, unpublished manuscripts, and correspondence, all of which had value for scientific and historical reasons.

Quite apart from the sheer size of the backlog of unanalyzed and unarchived material was the underlying fact that it had all been collected and documented under chaotic conditions and using inconsistent or localized systems for naming and description. By the end of the seventeenth century, with scholars traveling around and writing to one another, it became expedient to produce newsletters and to start having meetings, the proceedings of which were published and eventually became the first scientific periodicals.[42] This did much to resolve the issues of communication, but, despite the increasing formality of botanical Latin texts, in the nineteenth century practitioners were still operating without a set of universally applicable or enforceable rules. Adherence to Linnaeus' recommendations, and those of other botanists, was the decision and responsibility of the individual. There were several professional organizations, committees, and institutions that hosted meetings, conferences, and symposia, at which issues in the field could be debated and discussed. Founded in London in 1788, the Linnean Society not only hosted lectures and debates, but began publishing its *Transactions of the Linnean Society of London* in 1791, which ran until 1955. Nearly a century later in 1836, the Botanical Society of Edinburgh (BSE) and the Botanical Society of London (BSL) both held inaugural meetings. Each group focused on annual lectures and programs as well as on field excursions and on collecting and trading specimens for personal herbaria. They worked in tandem to produce a Flora of Britain and by 1850 had some 250 members, of which a large percentage were medical practitioners for whom membership in the Linnean Society was not feasible. They had also by this time produced the *London Catalogue of British Plants*, which helped to stabilize nomenclature of local flora.[43] Eventually, they combined to form the current Botanical Society of Britain & Ireland (BSBI). In India, the Nawab of Oudh, Saadat Ali Khan, established a royal botanical garden in 1789, which developed over time into the current National Botanical Research Institute (CSIR-NBRI). Under British rule, the garden, named *Sikandar Bagh* after one of Nawab Wajid Ali Shah's queens, hosted British horticulturists such Dr. J. Cameron, Dr. E. Bonavia, and G. M. Ridley. In the twentieth century, Indian plant scientists Prof. Birbal Sahni and Dr. S. K. Mukherjee expanded the garden, and in 1953 it was renamed the National Botanic Garden. By the late 1970s it had become a research institute. The Botanical Society of South Africa (Bot Soc SA) was founded in 1913, together with the Kirstenbosch National Botanical Garden. Two years later in 1915 they began publishing the *Journal of the Botanical Society of South Africa*, which focused on the garden's projects and research as well as the flora of South Africa. As is the case for many such institutions, the Botanical Institute of

Barcelona (IBB), founded in 1934, began with Francesc Martorell i Peña (1812–78), who had left his estate to the Barcelona City Council. Consisting of his natural history collection, library, and financial resources, it formed the Natural Sciences Museum, also called the Martorell Museum, which opened in 1882 and was the first public museum in the city.[44] These are just a few examples but they demonstrate that in general, by the late nineteenth/early twentieth century, while botanical institutes and organizations were popping up all over the world, their scope was often limited to the local flora. There was no governing body that ensured that plant scientists were all working under the same guidelines and assumptions and while meetings of naturalists were beginning to take place, interest in an international congress began to grow. In 1864, the first International Botanical Congress took place in Brussels, where a horticultural exhibition was being hosted. At the second congress in Amsterdam, the first formal call for standardized nomenclature was put forth by the German botanist Karl Koch. This was revisited at the third congress in London, where it was agreed that a decision would be made at the fourth congress, a conservative approach to acting on proposals that persists to this day.[45]

In 1867, French botanist Alphonse de Candolle (1806–93) published an early version of what would later become the International Code of Nomenclature. As motivation, he cited the historical "chaos" surrounding botanical nomenclature and descriptions as well as the weakening (*est ébranlée*) of the community's interest in strict adherence to the system laid out by Linnaeus.[46] According to him, the year prior in 1866 at the international meeting of botanists in London, "M. Charles Koch" (Karl Koch) had demanded that the botanical congress be employed to examine lingering questions about nomenclature and proposals to reduce the number of synonyms, that is, to prune everything down to the 1:1 ratio between plant and name recommended by Linnaeus.[47] This suggests that Koch's proposal from the second congress in 1865 had not been taken seriously, that he was frustrated enough to demand action, and that de Candolle took on the challenge ahead of the next annual meeting. His recommended "code" consists of several chapters or sections, each consisting of a number of articles that function as axioms and rules for naming, classifying, and describing plants. In particular, Article 6 in section 1 dictates that scientific names (*noms scientifiques*) must be in Latin and that where a name originates in a different language, it must be Latinized by declining the words as though they were Latin, by adding case endings to it (*ils prennent des désinences latines*).[48] Similarly, in Article 67, section 7, he reiterates that modern languages should be avoided for nomenclature and that they should also avoid using surnames (*ils évitent de se servir de ces derniers noms*),[49] unless they are commonplace and quite clear in meaning. In Article 68, he states that lovers of science ought to be opposed to the use of new nomenclature in a modern language unless that name is derived from a Latin name or can easily be modified to have a Latin form.[50]

The *Société botanique de France*, founded in 1854, hosted the fourth congress in Paris in 1867, the same year de Candolle's code was published. The primary goal of the Paris congress was to decide on rules for nomenclature, which were based on de Candolle's. Despite this, the rules were not enforced, only strongly recommended, until the 1905 Vienna congress, when formal Rules of Nomenclature were not only established but required for publication. It was at this same congress that French was established as the official language for congresses, while Latin was required, though not enforced, for descriptions beginning in 1908. The Vienna Code was published in 1906 as *Règles Internationales de la Nomenclature Botanique Adoptées par le Congrès International de Botanique de Vienne 1905*.[51] Article 7 of the 1912 edition[52] states that "scientific names are in Latin for all groups" and Article 36 states that as of January 1, 1908, "the publication of names of new groups of recent plants will be valid only when they are accompanied by a latin diagnosis."[53] In addition to these changes, the informal Kew Rule was abolished as contrary to the new established code. Named after the Royal Botanic Gardens, Kew, the Kew Rule had persisted since 1877 and allowed a species to be transferred to another genus without retaining its original epithet. That same year, Henry Trimen (1843–96), a British botanist who worked primarily in Sri Lanka and was a member of the Edinburgh Botanical Society, responded to a letter to the *Journal of Botany* by Alphonse de Candolle regarding the issue of synonyms. In his letter, de Candolle had argued that an author was at liberty to retain, or even transfer from another genus, an older specific epithet in favor of new ones when a species was transferred to another genus. Trimen responded that "to do this is to needlessly create in each case another synonym."[54] With the abolition of this practice, the principle of priority became commonplace, according to which the latest recognized and accepted name of a taxon took precedence. The Botanical Society of America (BSA), founded in 1893, decided at their 1914 Atlanta meeting to form an official publication, *The American Journal of Botany*, in conjunction with the Brooklyn Botanic Gardens, citing the need for "increased facilities for the prompt publication of the results of botanical investigation."[55] Unlike their European counterparts, the BSA did not seem to be concerned with the language of expression or with the institution of rules for naming and describing plants. However, at the 1930 Cambridge International Botanical Congress (IBC), it was established that beginning in 1932, Latin must be used for plant descriptions. In other words, they intended to begin enforcing the Latin requirement from the 1905 congress, which officially went into effect in 1908.[56] Although the Latin requirement officially began in 1932, the Cambridge Code was not published until 1935, and as a result, is alternately referred to as the 1935 Rule, the Latin Requirement, or the Cambridge Rule. According to Article 38 of this rule, the names of new botanical organisms "are considered as validly published only when they are accompanied by Latin diagnosis."[57] Prior to this, the use of Latin in botany was already ubiquitous, but the 1935 Latin Requirement was an attempt to ensure the diagnostic features of plants could be understood by anyone trained in the field. With its institution, botanical Latin instantly became a highly specialized and required tool for communication between botanists around the world.

Botanical Latin Post-1935: Debating the Latin Requirement

By 1935, the obligatory components for publication of a botanical taxon such as a family, a genus, a subgenus, or species had become:

1. The name of the group of organisms, a "binary epithet" or "binary combination."[58] This consists of the generic name, which is always capitalized, and the adjectival specific name, which is not capitalized.
2. The diagnosis, consisting of a short technical phrase, which must distinguish the taxon from others at the same botanical level. It is, basically, "a brief statement of the distinguishing features of an organism."[59]
3. The description, detailing the non-diagnostic defining properties of the organism.

Of these, the Code dictated only the formation of the name. For the diagnosis and description, there were no set rules for composition apart from the use of Latin. It did not take long for concerns to arise about the practicalities of this requirement. In 1949, in discussing Article 25, which required that the names of genera be always listed in the singular, members of the Central Committee on Nomenclature of the American Society of Plant Taxonomists argued to emend it to allow for the plural. They noted that the use of the plural did not imply multiple genera, because in Latin, the nominative plural case ending is different from the nominative singular and since Latin was the required language for nomenclature, it was "necessary to accept also the inflections of that tongue."[60] Another proposed emendation might strike the modern practitioner as needlessly exacting: the proponents suggested for Rec. XLIV that when a name is a compound of two or more words in either Greek or Latin, in Greek, the letter *o* be used to join the term, as in *Odontoglossum*, while in Latin, the letter *i* be used, for example, *cruciformis*. They argued that the current rule was "hopelessly inaccurate philologically."[61] What both of these proposals indicate is that the intricacies of classical Greek and Latin were already lost on many practitioners. It is the business of trained classical scholars, not botanists, to know that both languages avoided certain combinations of consonants, preferring either to insert a vowel or to contract them into double consonants. While not a criticism of the Latin requirement, these proposals show that the assumption that a trained botanist would know these details and devise names accordingly was incorrect.

Just one year later, in 1950, however, dissatisfaction with the Latin language was already developing. At the Stockholm congress, two concerns were raised, the first with regard to the names of plants cultivated as hybrids or through mutation and other similar processes. A special committee consisting of "representatives of horticulture from several countries" replaced Article 35 with a new version according to which such hybrids and mutations would "receive fancy epithets preferably in common language and markedly different from the Latin epithets of species or varieties."[62] The second issue directly addressed Latin diagnoses. The contingent of algologists unanimously voted to reject the Latin requirement for new species of algae, arguing instead for diagnoses in two languages, of which one must be English, German, or French. The first language was not specified and could have been any modern language. Despite being decades ahead of the debate and anticipating the eventual removal of the Latin requirement, this proposal was rejected.[63] Recalling the 1949 proposals of the American Society of Plant Taxonomists, more philological concerns were raised, this time with regard to the correct spelling of Latin names. The report states that there was a "vociferous wrangle" about whether or not Latin stems can end in a vowel and whether the connecting vowels of Rec. XLIV are actually connecting vowels or just stem endings. This was left up to the editorial committee to make a decision, which, the report concludes, "must obviously secure expert philological help in its deliberations."[64] These kinds of tedious debates about the minutiae of Latin grammar, spelling, and syntax continued for the next few decades, simmering in the years immediately following a congress and boiling over in the year or so immediately preceding an upcoming congress.

In the final decade of the twentieth century, however, the debate over the Latin requirement began ramping up, with several practitioners arguing that it should be retained. Kabuye (1990) appealed to the untranslated backlog of older botanical Latin literature, claiming that by learning Latin, botanists can understand the names, diagnoses, and descriptions and pass that knowledge on to future botanists.[65] This sentiment was commonplace and evokes the rationale given by William Stearn for his 1966 monograph, *Botanical Latin*, which explained the basics of Latin syntax, grammar, and vocabulary for practitioners who needed to compose names and diagnoses in Latin. In much the same way, the devolution of botanical Latin between Stearn's monograph and the 2012 Melbourne Code provided the basis for Emma Short and Alex George's 2013 monograph, *A Primer of Botanical Latin*, the main point of which is to school botanists in how to compose and read botanical Latin. Other practitioners such as Filgueiras (1997) argued that botanical Latin is stable, static, and easily understood.[66] Filgueiras claimed that it was no more or less difficult for a modern botanist to read and understand Linnaeus in the original Latin than for one of his contemporaries and that this is not the case for modern languages. The irony of this is that while Filgueiras is correct to question the stability of English, which has now superseded Latin for botanical texts, it does not follow that Latin is more intelligible. His argument is that Latin is no longer spoken, but has been preserved in the Code and therefore is no longer dynamic the way a modern spoken language is. This is an astute observation, since it could level the playing field for practitioners around the world. It is easier to learn the basics of a static language than to have to master a constantly evolving modern language that is rife with political and cultural norms and assumptions. Retreat into a language with no such dangers makes sense, but avoiding mistakes in composition or in translation is not solved by the retention of the Latin requirement. Modern terms in the vernacular still must be Latinized and as the early critiques of the Latin requirement show, it is a far more daunting task to master the basics of Latin than is reasonable to expect of a busy practitioner.

Other botanists were simultaneously arguing for the removal of the Latin requirement. Kostermans (1990) noted how few taxonomists were trained in Latin and the effect this lack of skill has had on the quality of formal diagnoses.[67] He pointed out how in order to compose their diagnoses in Latin, many taxonomists now had to depend on a small number of Latinists for assistance. There is very little overlap between plant scientists and classical scholars and the results of such collaborations were rarely more than middling and second-rate in quality. In the absence of a Latinist to oversee or check their work, taxonomists resorted to producing "language-faulty diagnoses, which, nonetheless, are good enough to be understood."[68] The question of course, is who understands even these imperfect diagnoses. Kostermans recognizes that the Latin composed by non-Latinists is unlikely to be correct, a fair assessment of a language as complicated as Latin and so different structurally from modern languages. The intelligibility and the "good enough to be understood" characterization come primarily from the technical vocabulary, which, even if it is incorrectly declined, will still be recognizable to taxonomists. Yet even this can be questioned. Chaudhri (1991) highlighted the naivete of Western botanists regarding the difficulty of making sense of botanical Latin for non-Western taxonomists, who, he argued, are blind to the "extent of hardships being faced by the students of taxonomy in our parts of the world." Chaudhri cites as an example the *Flora Iranica* of Karl Rechinger, whose Latin descriptions he calls unreadable and evidence that Latin as a tool for communication is ultimately ineffective for those in non-Western countries.[69]

The *Flora Iranica* consists of 176 volumes from 1963 on, with the last two published posthumously. Each volume focuses on one particular family of genera and species in Iran, Kurdistan, Afghanistan, Pakistan, Turkmenistan, Azerbaijan, and the Iranian Plateau. Rechinger's Latin diagnoses are, in contrast to Kosterman's concerns about quality, lengthy and robust. In volume 153 (1986), he describes six new species of *Gagea*, a genus of flowers in the Liliaceae family, in very competent Latin with the confidence of one who has either been trained in the language or has been composing diagnoses for so long it has become part of his skillset. For *G. anonyma* Rech. f., spec. nova (a new species), the formal diagnosis takes up 18 lines, following the format already standardized in the nineteenth century, describing parts of the flower with strings of ablatives. He begins with the bulb, which is 8–12mm in diameter and ovoid in shape, with thin, gray, subfibrous (somewhat fibrous) tunics (coatings), with a *collum* (neck) that is either very short, rudimentary, indistinct and dissected, or nonexistent, sending out subterranean "suckers" 1–2mm long, orange-colored when dry, and ending in little, subglobular (roundish) bulbils 2–3mm in diameter. The leaves and tepals (petal-like structures in the perianth, the sterile part of the flower), meanwhile, are semi-cylindrical, sulcate (grooved) on the top, and grey-green when dry. He also describes the caulis, the pedicels, the inflorescences, the capsules, and the leaves and tepals in greater detail. Even translated into English, with parenthetical explanations, this diagnosis, while it accomplishes the goal of distinguishing this species from others, is daunting. The technical vocabulary should not be a source of confusion for those trained in the field, yet even then,

Rechinger uses terms that have an unclear application, such as *surculus*, which means "sucker," referring to the shoots that a plant produces at its base. Used in the diagnosis of a bulb, however, one must surmise what the term *surculus* refers to. Here, it either means the roots or the adventitious roots, which form like suckers on the emerging stem of a bulb. Similarly, by using *collum* (neck) to describe the bulb, he is referring to the basal plate.

It is telling that a trained botanist like Chaudhri would find this entry unintelligible, since the terminology is tricky even in English. Chaudhri (1992) aimed his critique at more than just the language, however. He also questioned the validity and quality of botanical descriptions produced by Westerners. Part of the process of learning and knowing is doing, and even apart from the issue of colonization, Chaudhri is frustrated that botanists from Asia and Africa have been denied the opportunity to describe the flora of their own countries, arguing that "the flora of this part of the world has to be studied by the taxonomists of these countries themselves."[70] The timing of this debate is significant. The Convention on Biological Diversity (CBD), initiated by the United Nations Environment Programme's Secretariat of the Convention on Biological Diversity, was established in 1992 in Nairobi. There had been a series of *ad hoc* United Nations working groups convened in 1988 and 1989 to discuss species extinction and threats to ecosystems, and to discuss an international convention, or set of standards and legal best practices, to ensure the "conservation and sustainable use of biological diversity."[71] Article 8 states that each contracting party should, "subject to its national legislation, respect, preserve and maintain knowledge, innovations and practices of indigenous and local communities."[72] For botany and other disciplines, this meant that it was now unethical and possibly illegal from an international standpoint, to obtain external funding and permits to collect and study flora in other countries without internal input. In particular, the tradition of researchers collecting species and naming them as though they were new, was increasingly being critiqued. Researchers looking to work in other countries had now to involve taxonomists from the country of origin of the organisms being studied.[73] The practicalities of this were in many situations unclear, since some host countries were suspicious of outsiders, and rightly so. Some suggestions included working with graduate students in home institutions and devising research projects that were collaborative and had clearly stated, viable outcomes that would benefit local communities, what the CBD called the "sharing of benefits."[74] The International Botanical Congress's continued insistence on Latin for the naming and description of new taxa created a problem for practitioners looking to adhere to the CBD's recommendations.

McNeill (1997) also pointed out the disparity, noting that simply speaking a first language with Latin roots provides a distinct advantage to practitioners.[75] A plant scientist from China, Russia, Indonesia, or even from a European country with less representation in the history of botany, should not have to become proficient in a language other than English, especially when that language has such a narrow range of application. He argues that the Latin requirement puts those with a native language with some connection to Latin at an advantage and that the ideal of a level playing field in the form of an "equal disadvantage" is an illusion. The relative stability of Latin compared with a modern language means little when the person reading it is completely unfamiliar with it. In the decades since this particular debate, technology has advanced in such a way that some of the concerns raised are less immediate than before. Artificial intelligence, digitization, and online databases have all to a certain extent ameliorated some of the issues highlighted. Yet they do not resolve the philosophical and ethical issues at stake. For example, in a 2006 entry for the new genus *Physeterostemon*, the authors explain its distinction by reference to the features of its flower parts and stem.[76] In contrast to Rechinger's style, this diagnosis is brief, consisting of only five lines. It uses the same ablative constructions, but begins with a short sentence in which the authors state that this is a new genus of uncertain placement regarding its tribe, a taxonomic rank between subfamily and genus: *Hoc genus novum quoad tribum incertae sedis*. The authors have cited a Latinist who has assisted them, which is clear even from this first sentence, since the main verb in the English translation ("it is"), which should be *est* in Latin, is elided, a common technique in classical literature, but not frequently seen in botanical Latin. The remainder of the diagnosis is also a full sentence with a conjugated verb and adverb, "it is easily diagnosed" (*facile diagnoscitur*). The ablative constructions all hang off of this verb: "from all of the other genera of the Melastomataceae" (*ab omnibus aliis generibus Melastomatacearum*); "by its hexameric and diplostemonous flowers" (*floribus hexameris et diplostemis*); "by a persistent calyx" (*calyce persistenti*), "by an inferior ovary" (*ovario infero*), and so on. Since this is a new genus, the authors must provide a type species, *Physeterostemon fiaschii*. The diagnosis for this type species is even shorter, consisting of only one line: "This new species chiefly differs from *P. jardimii* by virtue of its non-glandular hypanthium and its glabrous ovary" (*Haec species nova a P. jardimii hypanthio eglanduloso et ovario glabro praecipue differt*.[77] This is yet another idiosyncrasy, since it was not common at this time to place the main verb in the last position of the sentence, considered neutral word order in antiquity. In fact, by the end of the twentieth century, as Rechinger's diagnoses show, it had become common to write in a style evocative of Kaspar Bauhin's several centuries before, to use only and exactly as many words as were needed to make the diagnosis. Since copulative verbs (such as "to be") in particular are often implied in Latin, they were regularly elided. However, apart from the elision in the diagnosis of the genus and the verb position in the species diagnosis, the entries are typical of modern botanical Latin and are proper, scientific diagnoses. Both are highly technical, making heavy use of the ablative case to form complex secondary constructions. For the species, brevity is preferred, since there is no need to state further information in the diagnosis of the species, which would be identified in the same manner as the genus. Since this new species needs to be diagnosed as novel, however, it needs to be distinguished from others, in this case, *P. jardimii*. By 2006, the use of secondary ablative constructions was merely an accepted convention for fulfilling the requirements for publication. The use of the ablative is a *de facto* requirement, not a stylistic choice. Botanists used the ablative in these ways because they did not know Latin well enough to compose a diagnosis in any other way. They would master using the ablative without actually knowing much Latin, but as even this skill became diluted, botanists began simply copying the ablative constructions they saw in other diagnoses, such as those in Rechinger, until eventually it became standardized. The result is a peculiar textual artifact that exists only because it was required for publication. The diagnosis of *Physeterostemon* is an idiosyncratic admixture of technical terminology, formatting requirements, syntactic conventions, and classical markers like the placement of the verb at the end of the sentence. The name is also inscrutable to anyone who does not know Greek. *Physeter* is Greek for sperm whale, while *-stemon* refers to the stamens. The accompanying illustration, in which the flower parts are dissected out, communicates this quite well, since the stamen has the shape of a sperm whale's head, but what really makes this clear is the English-language etymology provided by the authors. They also provide an English translation of the diagnosis, which, while far more legible to practitioners in the field, is not exact and leaves out some of the information in the Latin. It omits the first sentence entirely and then, when the series of ablatives gets too lengthy, the translation switches to a different structure entirely, making the androecium the subject of an entirely new verb ("the androecium consists of . . ."). For the calyx, an extra identifying feature has been added so that it is now described as "toothed and persistent calyx," where the Latin has only *persistenti*. This leads one to wonder whether the Latin came first or the English and which is the more precise. For practitioners, the answer to this is simple: the English translation states mostly the same information as the Latin in a format that is familiar to other botanists. The Latin was most likely a translation from the English and was included only because it was required. More concerning is the issue of reliability, since an author who is unfamiliar with Latin has to trust the Latinist to correctly state the defining features of the organism, while the Latinist must assume that any mistakes have been caught by the authors. Ideally, the Latinist was someone who, like Stearn, works frequently with botanists and has learned how they communicate with one another, but this was not always possible.

The Latin Requirement was still being debated at this time and Kosterman's point about "good enough" diagnoses was becoming more and more clear. Botanical Latin had devolved considerably in the twentieth century and the international community was becoming more and more frustrated. Despite this, it was not until 2011, at the Eighteenth International Botanical Congress in Melbourne, that the decision was made to no longer require Latin for the publication of technical, botanical diagnoses of new organisms. The result of the Melbourne Congress was the new International Code of Nomenclature for algae, fungi, and plants (ICN), put into effect on January 1, 2012 and generally referred to as the Melbourne Code.[78] According to this Code and those subsequent to it, the names of organisms must still be devised in Latin and must still adhere to Linnaeus' binary rules of nomenclature. However, diagnoses may now be published in either Latin or English. Article 39.2 states that the name of a new organism must have a description or diagnosis in either Latin or English, or must refer to a previously published description or diagnosis in Latin or English. This was an attempt to placate both sides of the debate, to allow those who argued for the intelligibility, stability, and accessibility of Latin to continue to use it, while allowing those who have no familiarity with Latin to use English instead. Yet, despite the vehemence of the arguments in support of Latin, the Melbourne Code effectively rendered botanical Latin obsolete apart from nomenclature. Validly published diagnoses are now almost exclusively in English and those written in Latin are relegated to the history of the discipline. On that level, the late-twentieth-century debate about botanical Latin found a tense resolution. On a deeper level, however, the replacement of one obsolete language for a current *lingua franca* as a requirement for diagnosing and describing plants in a valid publication and for recognition by a specific, elite group of peers, does not resolve the ethical and philosophical issues associated with the Latin language. That recognition involves authorship and therefore authority, "discovery," and a false sense of ownership over plants. These issues are still embedded in Latin nomenclature, but they will persist as long as intelligibility and knowability continue to be dependent on natural languages.

Conclusion

The preamble for the Convention on Biological Diversity expressly recognizes the "close and traditional dependence"[79] that indigenous and other groups have on the very same natural resources being collected, examined, studied, preserved, named, and dispatched overseas by practitioners in the field of botany. Switching from one colonist language to another does nothing to address or assuage the harm caused by taking these resources away from local communities. Even the assumption that English is universally intelligible enough to justify its requirement for valid publication is problematic; it takes little more than the recognition of a single group of trained botanists for whom English is neither their first language nor their preferred language of scientific expression to serve as a counterpoint. The truth is that English is convenient for practitioners whose nations have colonized others and who see their own traditions as superior. There is no philosophical reason, nor any valid scientific reason, why a group of botanists cannot simply refuse to describe plants in English. Recognition by the IBC is about professional acceptance and assimilation into an exclusive club of individuals and institutions that promote specific ways of knowing, discovering, naming, and describing plants. They are working with a very narrow conception of intelligibility, according to which the extent to which marginalized groups can attain it is completely irrelevant at best and purposefully ignored and sidelined at worst. Within the framework of transformation methodology, these debates about botanical Latin become a reception sphere where the reference is more than just one author-text datum. It is a complex tension between the legacy of colonialist botanical expeditions and the formalization of botanical Latin and nomenclature. The former, aside from the subjugation and enslavement of indigenous peoples worldwide and the suppression of their knowledge and traditions, produced vast collections of specimens, seeds, narrative accounts, notes, observations, and more, all of which is knowledge defined by the methods by which it was attained. The latter involves a literal transformation of the ancient tradition of naming and describing plants in Latin into a rigid rule book that overtly excludes anyone who does not follow its regulations. It is tempting to treat both of these phenomena separately, as mere temporal points on the continuum of botanical Latin, but the emergence of the International Botanical

Congress and the development of the Code did not happen in a vacuum. The centuries of botanical colonialism may have brought prestige and acclaim to European and European-American botanists, and their herbaria and botanical gardens may have been enhanced, but indigenous knowledge of those same plants, with their numerous common names, was an ever-present reality that bothered the consciences of some and irritated the haughty condescension of others. The development of the Code was just as much about unintentional silencing as it was about ensuring universal intelligibility amongst a select group of practitioners.

5

Current Debates and Visual Communication

An unexpected effect of historical botanical colonization was the development of the International Code of Nomenclature (ICN) and the introduction of the Latin rule for scientific descriptions and diagnoses of plants. Together, they occupy their own places on the continuum of botanical Latin's history. Similarly, the twentieth-century backlash against the 1935 Latin rule directly affected another crucial point on this continuum: the rejection of the Latin requirement in 2011 at the Melbourne International Botanical Congress (IBC). The revised rule, effective January 1, 2012, dictates that diagnoses and descriptions must be composed in either English or Latin. As a result, almost no new published taxa have diagnoses in Latin, and English has become the standard. The revision has accomplished much to make current botanical literature accessible and intelligible to more practitioners, while also being more in keeping with Linnaeus' definition of a botanist as one who knows how to assign similar names that are intelligible to everyone to similar plants.[1] The IBC continues to regulate the formation of botanical nomenclature, which must still be in Latin. Modern, post-2011 botanical Latin has been reduced almost entirely to nomenclature and to the historical backlog of untranslated botanical texts. As of 2025, the Madrid Code (Article 32.1) dictates that the names of taxa must be composed of Latin characters, with a few exceptions.[2] This means that any name originating in a language with a non-Roman alphabet must in effect be Latinized. It is not sufficient to transcribe non-Roman languages by replacing the characters with their Latin counterparts, because the IBC also insists on retaining the binomial format for epithets, as conceived by Linnaeus. In this model, binomial epithets involve a genus name followed by a species name, either a noun in apposition or an adjective that modifies the genus name, in the exact way that Latin adjectives are declined and modify nouns, matching them in gender, case, and number. If a genus name is feminine, nominative plural, an adjectival species name must also have these features. For a name transcribed from a language not inflected like Latin or Greek, it will not be possible to match the two names unless it is first

Latinized and assigned gender, case, and number. The IBC occurs every six years and the 2011 congress was followed by the 2017 congress in Shenzhen, China, which produced the 2018 Shenzhen Code, with another in 2024 in Madrid, which resulted in the 2025 Madrid Code. In the lead up to both congresses, there was significant debate about the continued use of Latin in nomenclature. The controversy has been focused on two main issues: the historical deprivileging of indigenous names and the prevalence of honorifics in the current, established nomenclature, which occasionally memorialize colonizers and slave-owners.[3] In addition, given the deposition of Latin as the *lingua franca*, there is a continued, glaring disconnect between the current nomenclatural requirements and the discipline's historical roots. Pliny the Elder and Carl Linnaeus both argued for Latin as a communicative tool useful for universal intelligibility, but in the twenty-first century, Latin botanical names have become unintelligible to most people.

Colonization in Nomenclature

The 2012 Melbourne Code (ICN) retained Latin for botanical nomenclature. However, more recently, this too has been critiqued. Smith and Figueiredo (2022) have argued that in many cases the use of personal names in nomenclature encodes the effects of colonialism.[4] Many botanical names are Latinized versions of colonizer names, a particularly egregious example of which is the genus name *Hibbertia*, commemorating George Hibbert (1757–1837), an English politician and merchant and a slave-owner. Hibbert opposed the Act for the Abolition of the Slave Trade of 1807, likely due to the fact that he was also the first chairman of the West India Dock Company. He was also an amateur botanist, a member of the Linnean Society, and the founder of the London Institution in 1805. Hibbert's legacy in the genus name implies that he should be credited with the discovery of species within it, such as *Hibbertia scandens*, formerly *H. volubilis*. This was the first species named after Hibbert in 1800 by Henry Cranke Andrews (*c.* 1759–1830) in *The Botanist's Repository for New and Rare Plants*. Andrews claimed that *H. volubilis* is from New Holland and was propagated by seed by numerous people in and around London in 1792. He decided to name it after Hibbert because "hitherto this tribe of Plants has not been given to the public by any truly established name." Local names for this plant, even Latinized, were unacceptable for Andrews, but Hibbert was worthy because of his "knowledge and fervor in botanical pursuits."[5] In fact, Hibbert primarily funded expeditions, such as to South Africa, where the Scottish gardener James Niven (1776–1827) was sent to collect specimens. There, around 1805 in Cape Province, Niven "discovered" several species of the genus *Protea* (sugarbushes), such as *Protea eximia*. Thiele et al. (2022) argue that names such as *Hibbertia* violate Article 51.1 of the 2018 ICN (Shenzhen Code), according to which, "a legitimate name must not be rejected merely because it, or its epithet, is inappropriate or disagreeable." Their argument is that the name *Hibbertia* is more than just inappropriate; it is offensive and therefore should be rejected. It also violates some of the claims made in the ICN preamble. Preamble 1 states both that, "biology requires a precise and simple system of nomenclature that is used in all countries," and "the purpose of giving a name to a taxonomic group is not to indicate its characters or history, but to supply a means of referring to it and to indicate its taxonomic rank." Meanwhile, Preamble 5 states that, "names contrary to a rule cannot be maintained."[6] If the name *Hibbertia* is in fact in violation of these rules, then it must be changed. There is also historical precedent for its rejection, since names like *Hibbertia* violate Linnaeus' own recommendations. The convention of naming plants after their "discoverer" was explicitly forbidden in Linnaeus' Aphorism #263:

> *Nomen Inventoris, vel alius cuiuscunque, in differentia non adhibeatur.*[7]
>
> The name of the discoverer, or of anyone else, should not be adopted in the differential [diagnosis].

Critics such as Mosyakin (2022a, 2022b) have claimed that renaming such genera creates a slippery slope, arguing that if *Hibbertia* must be renamed, then so too must other genera such as *Banksia*, named after Joseph Banks (1743–1820), the English naturalist and botanist and long-time president of the Royal Society. Banks traveled to Newfoundland, New Zealand, Australia, and Brazil, and introduced to the Western world a number of plant species, bringing 30,000 specimens back and cataloging some 1,400 of them. Mosyakin also suggests that the genera *Darwinia*, *Linnaea*, and even *Aristotelia* should all be renamed on the same grounds, that their namesakes had questionable ideas by modern standards, stating that the elimination of names that memorialize or honor racism, imperialism, and slavery should be done "systemically and consistently, without any exceptions."[8] Thiele et al. (2022) counter that there is a distinction between Hibbert and Banks. The former was an anti-abolition slave-owner, while the latter was simply complicit in the colonial projects of the English Parliament and Crown. It is therefore an issue of degrees of harm and plausible deniability. There is a continuum of injustice caused by these individuals, which is difficult to parse, since "societies everywhere manage shades of grey very well, and do not collapse because some decisions are clearer than others."[9] Other questionable honorifics include *Kalanchoe salazarii* Raym.-Hamet, a species native to Angola and named after the authoritarian Portuguese statesman António de Oliveira Salazar (1899–1970); *Gardenia leopoldiana* De Wild. & T. Durand, a species native to Cameroon, Gabon, and Democratic Republic of Congo, named after Leopold II of Belgium (1835–1907); and *Tragia rhodesiae* Pax, a species native to Malawi, Zambia, and Zimbabwe, named after Cecil John Rhodes (1853–1902), the prime minister of Cape Colony, 1890–6.

Botanists expressing these concerns are not arguing for the wholesale dismissal of such honorifics in botanical nomenclature, but rather for their careful review by specialist Permanent Nomenclature Committees under the auspices of the General Committee of the IBC. At the 2024 Madrid IBC, these issues were discussed and voted on, with the results published in the 2025 Code. Nicholas Turland (2025) summarized the most significant changes, including: (1) changing the spelling of very specific offensive names;[10] (2) an additional phrase added to the existing Article 56.1, that any derogatory name can be proposed for rejection; and (3) a strong admonition to authors to just avoid potentially offensive names in the first place.[11] The new Article 56.1 recommendation contains the phrasing: "Any name that would cause a disadvantageous nomenclatural change (Art. 14.1) or that is derogatory to a group of people (Art. 51.2) may be proposed for rejection."[12] The decision to strongly advise authors against the use of offensive names was proposed by Mosyakin (2021), who argued that the Code should not retroactively judge the entire corpus and history of plant nomenclature and that only *ad hoc* judgments can be made in individual cases.[13] It is easier to warn authors that some names might be "inappropriate, disagreeable, offensive, or unacceptable by any national, ethnic, cultural, or other groups."[14] While these are important steps toward resolving the issues inherent in botanical nomenclature, they do not address the underlying moral and philosophical issues at stake. The practicalities of re-naming such taxa are not mere exigencies; there are serious moral—not scientific—problems with botanical nomenclature that such measures do not address. Individual honorifics are one thing, but the whole of Latin botanical nomenclature is a relic of colonialism.

A second, related issue has therefore arisen, regarding indigenous plant names. Wright and Gillman (2022) have argued that the Latin names of some plants should be replaced with their indigenous names, noting that, "names are important to people; they affirm a sense of place, belonging and history."[15] They therefore recommend a formal process for replacing current validly published names with their earlier, indigenous names, under the assumption that "chronological precedence, in discovery, naming and usage, should be afforded priority." This, too, is significant for practitioners in the field to acknowledge, but the solution offered is less than ideal. Given that there are in some cases multiple known indigenous names for plants, the authors suggested that indigenous groups would need to arrive at a consensus on exactly one indigenous name to replace the Latin nomenclature, with a focus on species names, although names at any taxonomic rank can theoretically be replaced.[16] In the absence of a consensus, the name was to remain unchanged. Apart from the immorality of placing the onus for selecting a prior indigenous name onto marginalized peoples who may have little to no interest in such debates, it is hardly a solution to an overarching problem to fall back on inaction. Another proposal that did not pass at the Madrid congress would have allowed for vernacular names in a protologue, the complete publication of a new species. The authors suggested that to Article 38 a recommendation could be added, encouraging authors to report indigenous and other vernacular names and, if possible, to use them in the formal scientific name.[17] This would circumvent the problem of invisible indigenous names in the future, but as Earp and Mosyakin (2024) note, such information was not prohibited in the 2018 Shenzhen Code (per Art. 6.13). Had this proposal gone through, taxonomists would simply have been encouraged to do what they are already able to do. At best, were more taxonomists to include such information in their protologues, it would begin to be seen by practitioners as scientifically valid information, not merely historical curiosities, and "this kind of information from the protologue might appropriately be used for various purposes related to plant taxonomy and nomenclature."[18] In other words, scientists might deign to consider traditional knowledge worthy of their notice. None of these proposals engage with the real issue, that knowing and preserving indigenous plant names in scientific publications does not erase the colonial nature of botanical nomenclature, which continues to be privileged over traditional names in the vernacular. Nor does it address the problem of what it means to know a plant.

Turner (2014) has an entire appendix with an extensive list of indigenous names compiled for a variety of plants native to western North America. The species in this list have long been known to the people who speak these languages. For bull kelp (*Nereocystis luetkeana*), Turner lists twenty-six tribes, some of which have more than one name for the plant, depending on whether it is the full plant or one of its parts. For example, *waaqaat*, which means "circumcised" or "extending down," is used by the Ditidaht tribe for the entire plant, while *lataabub*, meaning "hair," is used for the bulb. Similarly, some names refer to the uses of bull kelp, for example, *tl'agaay*, from *tl'aga* ("to soak"), used by the Skidegate Haida for fishing line made of bull kelp.[19] Any refusal to re-privilege such names is a tacit acceptance of a particular, Western idea of science and knowing. This brings us back to the issue of intelligibility. In the 2006 entry for *Physeterostemon* discussed in Chapter 4, it is possible to ask to what extent the intelligibility of the Latin name and diagnosis promotes knowledge of the plant. Simply being able to decipher the literal meaning of the words does not mean that anyone besides a botanist will have any idea what the genus looks like, how it grows, where it is from, or how it has traditionally been used. Despite the aims of the 1935 IRBN, access to the specialized tool for communication that is "botanical Latin" was denied to many botanists and the field continues to struggle with its colonial past.

The Role of the Visual in Intelligibility: Scientific Specimens and Illustrations

For the most part, these debates between practitioners, most of whom are taxonomists, have been limited by the parameters of text and language, whether verbal or written. However, the ideal of universal intelligibility and its related concepts of discoverability and knowability, extends beyond the confines of text or speech-based languages to visual communication. Botanical specimens, illustrations, and even nature prints (made from specimens) all convey information, whether as supplements to scientific texts or as decoration, fine art, commercial catalogs, or as an integral component of "folk" or indigenous practices. Botanical illustrations have their own unique history, distinct from specimens and nature prints. In the long history of botanical texts, illustrations were usually the only visuals one could expect to find and, unlike specimens and nature prints, there is considerable interpretation involved in their production. Karen Reeds (2006) discusses the fifteenth- and sixteenth-century phenomenon of the nature print and its lack of representation in scholarship on illustrations, calling it mere "child's play."[20] Illustrations, on the other hand, are far more complex; they communicate both "objective data" about a plant and the ideals and beliefs of their creators. Simultaneously, they reflect the ideals embedded in their accompanying texts.

According to Article 40.4 of the 2018 Shenzhen Code, in order for the publication of a new taxon to be valid, it must include with the diagnosis and description a specimen of the plant in question.[21] Prior to 2007, it was possible to use an illustration, although these are now only allowed for microscopic algae or microfungi, not including fossils, for which the preservation of specimens may be technically difficult.[22] Specifically, whether specimen or illustration, the image accompanying the text must be what is called a type. If one is publishing the name of a new genus, the authors must include a type species, since every genus has at least one species. The illustrations and specimens in scientific publications are therefore literally specific. With this insistence on visual information, the scientific community has recognized that mere text is sometimes insufficient to make a plant knowable to the broader community and that one's words must be supplemented by a visual. The placement of a scientific illustration, especially immediately after the scientific name, also draws the reader's attention to the image as a holotype of the new taxon being described.[23] The image is a "visual philology" that aligns nomenclature with organism. Latour (1990) called this kind of image an "immutable mobile"[24] because it functions as a "mobile representative of the relatively immobile specimen."[25] This convention, which assumes a symbiotic relationship between text and image, is evidence that the ideal of universal intelligibility is unattainable by text alone and is not possible in any particular natural language. In what follows, the history of botanical illustrations is briefly considered, followed by a number of case studies in which the information conveyed in early scientific botanical texts is analyzed in the context of the premise that text and image must work in conjunction in order to attain intelligibility and knowability.[26]

The earliest known botanical illustrations in the Western tradition are usually attributed to the botanist Crateuas, physician to Mithridates IV Eupator, king of Pontus in Asia Minor, and author of the now-lost *Rhizotomikon*, considered the earliest illustrated herbal.[27] According to Pliny the Elder (*HN* 25.8), there were others who illustrated the plants about which they wrote, including Dionysius and Metrodorus, but of these, only Crateuas' have survived, about eleven of them having been copied into three separate codices of Dioscorides' *De Materia Medica*. Pliny's criticism of plant illustrations may seem to contradict his methodology,[28] but it makes sense if one considers that such illustrations were not realistic but highly stylized, much like the illustration for frankincense in the tenth-century Julia Anicia codex of Dioscorides (see Figure 5.1). The purpose of such images seems to have been to provide something visual for the eye to rest on, without concern for identification in the wild. The style is abstract, simplistic,

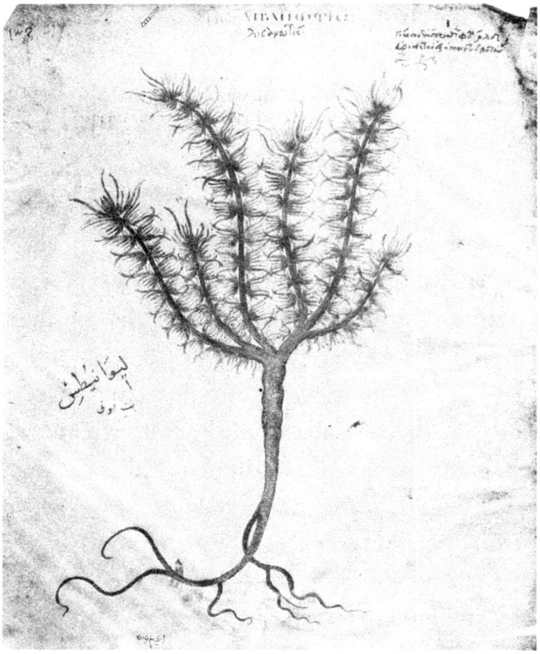

Figure 5.1 *Frankincense*, Dioscorides. Julia Anicia Codex, tenth century. Credit: M0006401. Courtesy: Wellcome Collection.

and devoid of identifying details. Despite this, and perhaps because it was attached to Dioscorides, whose herbal was copied over and over, it persisted throughout the medieval herbal tradition.[29] The reasons for this enduring lack of detail and inexactness are varied, but the intended audience of a herbal was usually a medical practitioner or herbalist with sufficient pre-existing knowledge of the herbs depicted. At the same time, some scribes with insufficient knowledge of herbs simply copied illustrations from one manuscript to another, increasingly abstracting them from the actual plant in nature. Despite this, ancient and medieval herbals were neither unscientific nor unsystematic. Referencing Lowden's (2002) concept of "visual knowledge" as uniformity in iconography derived from experience within a particular artistic tradition,[30] Griebeler (2024) argues that the botanical illustrations in these texts "aimed to create and communicate visual knowledge about plants."[31] The style of illustration associated with medieval herbals is not simply the product of lazy copying or a lack of artistic skill. It is the visual or iconographic result of the application of the accumulation of centuries of knowledge and expectations with regard to how certain plants should be depicted. This is evident in some of the illustrations of

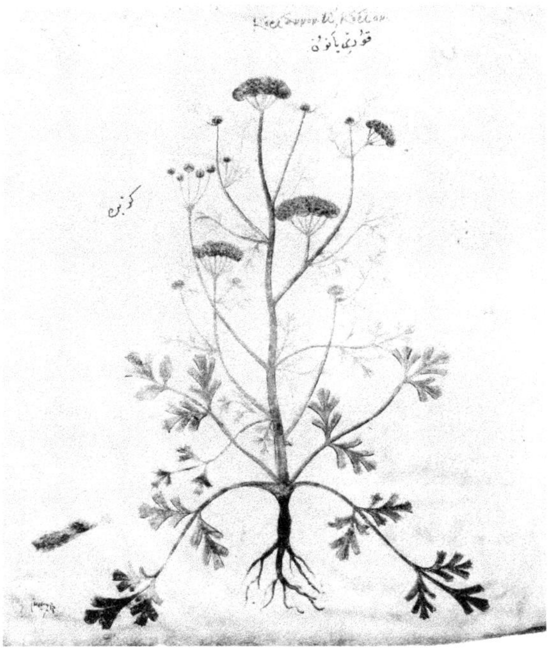

Figure 5.2 *Coriander*, Dioscorides. Julia Anicia Codex, tenth century. Credit: M0006403. Courtesy: Wellcome Collection.

Dioscorides such as coriander, a well-known and well-used herb that was depicted in much the same manner across the herbal tradition (see Figure 5.2). Despite this, there were other concerns with the tradition. Pliny had also complained about the secrecy surrounding plant knowledge and the fact that practitioners purposefully misled their readers, by using pseudonyms or by indulging in hyperbole in their descriptions of plants. This continued even in the Renaissance, where we can identify a certain indulgence in alchemical plants with magical properties and accompanying fantastical illustrations, the mandrake being one of the most common (see Figure 5.3).[32] Pächt (1950) provides an overview of the "slightly more scientific" natural historical works that proliferated in Italy as early as the thirteenth century, herbals included. He argues that the "regeneration" of ancient herbals as useful compendiums in Italy was due in part to contact with the Arabic tradition that had preserved the Greek scientific and medicinal traditions. In order to render herbals useful again as handbooks after centuries of corruption, illustrators began referring to nature.[33]

Illustrations were inconsistently used in European botanical texts after the sixteenth century, in part because they had been associated with the herbal

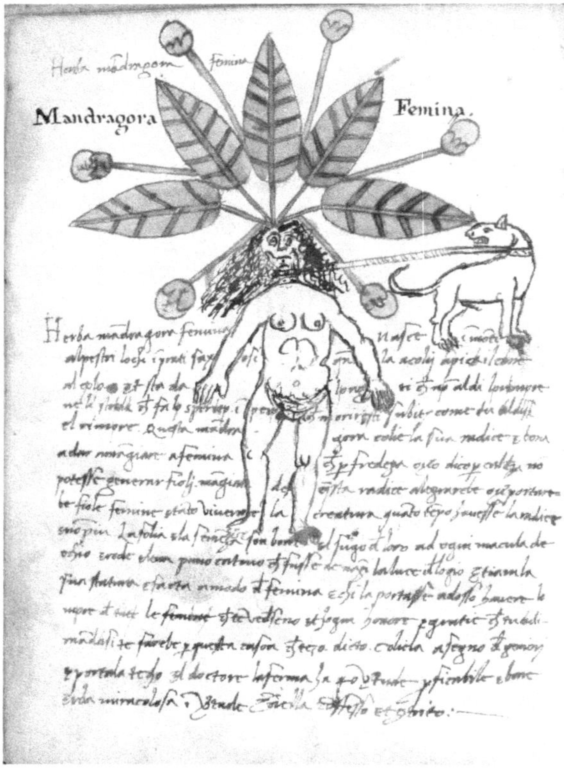

Figure 5.3 *Mandragora femina*. Penn LJS 419, *Erbario*, fourteenth century, Italian. This illustration depicts the myth surrounding correct harvesting of a mandrake. Credit: Kislak Center for Special Collections.

tradition for so long. However, the value of scientific illustrations, which often have a particular style and depict both a full view of the plant and its identifying parts, was rediscovered in the sixteenth century. The early-twentieth-century historian Agnes Arber was for a long time the authority on the medieval herbal tradition. In the essay "From Medieval Herbalism to the Birth of Modern Botany," Arber focused on the effect on the development of botany as a science of the illustrations of plants in printed books.[34] For example, it was not uncommon for printers to reuse woodcuts of one plant to illustrate another, to save money. At the same time, the same name is often used for different plants, even those from different genera, the resulting confusion of which played a role in the development of Linnaeus' binomials. Arber's work is historical, tracing the progression of herbals over the course of two centuries, from 1470 to 1670, all the way to botany as a modern science. Outside of the world of book scholarship, in the scientific community, Arber is the precursor to Martin Kemp, whose 1990 *The Science of Art* fleshes out the proposition that illustrations have a function in scientific identifications, descriptions, and diagnoses of taxa. Botanical illustrations are now commonplace and can be compared to earlier traditions in terms of how they relate to discoverability, knowability, and universal intelligibility. Griebeler (2024) argues that the totality of knowledge about the observable, natural world can be "conveyed and transmitted primarily through visual, depictive means."[35] Although their work is focused more broadly on science in general, Daston (2008) notes that since the sixteenth century, botanical illustrations have been focused on idealized images that typify a particular species.[36] They are not meant only to represent the individual specimen used for the image, but are "schematic"; they convey visual information that allows the "reader" to understand the identifying characteristics of the taxon. This has not been particularly problematic in botany, however, since the connection between a living organism in the real world and the illustration is easier to establish. In contrast with Griebeler, Daston considers scientific research to be fundamentally tied to the reading of texts, despite the "bibliophobic rhetoric" that has characterized the sciences. Scientific research is both observation of physical objects and reading or writing about them. The process of scientific observation and learning is actually a triangulation between images, words, and physical objects. In other words, "to read means also to see."[37] The unstated implication of this claim is that the opposite is also true and "to see is also to read," in the sense that reading leads to understanding and knowledge. Both approaches demonstrate a turn toward a more understanding, less harshly critical analysis of the value of the history of herbals and botanical illustration. Every method of knowing plants, whether oral, textual, visual, or experience-based, is inherently limited by the cultural and linguistic contexts of its propagators. Moreover, no method of knowing is comprehensive. Everything knowable exists on a continuum with no defined terminus. There are no hard boundaries and the limits are inherently fuzzy.

From late antiquity until the sixteenth century, the art of botanical illustration languished as an afterthought.[38] This lasted until the publication of Otto Brunfels' 1530 *Herbarum Vivae Eicones* (Living Images of Herbs). Brunfels' publisher, Johannes Schott, employed the artist Hans Weiditz to produce realistic illustrations of plants throughout Germany. Weiditz's procedure involved producing watercolor paintings of the plants, from which line drawings were made and transferred in reverse to woodblocks. In turn, a skilled engraver cut the woodblocks.[39] The original watercolors were lost until the 1930s, at which point they were reproduced under the title *Pflanzenaquarelle* (Watercolors of

Plants).[40] The value of these illustrations was mostly lost on Brunfels, which is ironic given that there were several plants that he could not positively identify, which he called "bare herbs" (*herbae nudae*). Brunfels' German readers might have already been familiar with the "bare herb" *Gauchblům*, for example,[41] but an international audience would have needed to consult the illustration. Nor was the purpose of the illustrations necessarily or solely about identification in the wild; they were helpful for clearing up confusion about their names and descriptions throughout history. Brunfels was engaged in a project of verification and his reliance both on specimens and the illustrations of Weiditz was about mapping the German names he knew to the Greek and Latin names from antiquity and the medieval eras. Weiditz could not have known ahead of time that Brunfels would have trouble identifying some of the plants. Brunfels' process had been to provide to Weiditz a list of herbs to find in the field, paint, and procure specimens. Brunfels then wrote about each plant as its illustration reached him from the field and we can assume that Weiditz painted them as he saw them.

So faithful were Weiditz's renderings that some historians of science have criticized their scientific value, echoing Pliny's complaint in the first century that illustrations only depict a plant in one stage of growth. In some cases, Weiditz faithfully reproduced wilted leaves, dead flowers, and other features of the plant that, while true to the individual specimen, were not particularly helpful for identifying the species in the wild. In the illustration of *Gauchblům*, Weiditz focused on portraying the flower shapes, including petals and stamens, flower buds, leaf shape and spread, the upright stalk, and the roots (see Figure 5.4). While realistic, there seems to be little artistic interpretation and the end result is something Rix (1981) calls "feeble and atypical."[42] Weiditz's illustrations are too detailed, showing a plant's imperfections. This criticism is far clearer in the 1937 *Pflanzenaquarelle* color reproductions than in the black-and-white woodcuts in the printed book of 1530. For example, plate XI of the reproductions shows three herbs, including *Cichorium Endivia L.* on the far right, which is depicted with browning, wilted leaves. From the original watercolor, it appears that while Weiditz might have been overly literal, it is also possible that the specimens from which he was working were the best he could find. In terms of the information conveyed in an illustration, this attention to seemingly unnecessary detail is not a serious problem, despite the criticisms; it is simply a different concept of what one considers important information to convey.

A more substantive critique might be aimed at Brunfels, who had refused to provide more than the briefest of descriptions of *Gauchblům* and the other

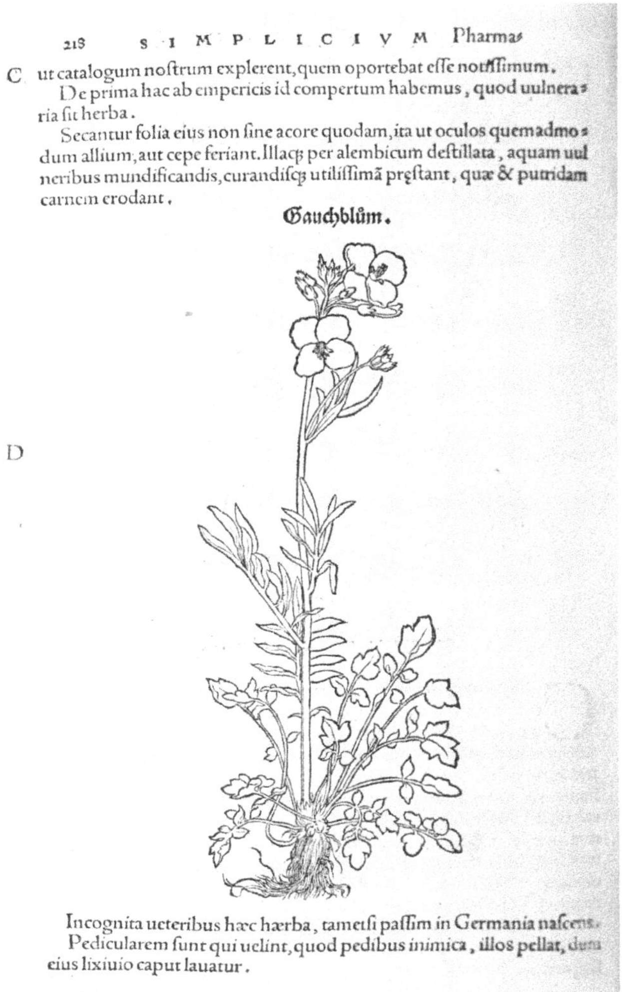
218 SIMPLICIUM Pharma-

C ut catalogum nostrum explerent, quem oportebat esse notissimum.
De prima hac ab empericis id compertum habemus, quod uulneraria sit herba.
Secantur folia eius non sine acore quodam, ita ut oculos quemadmodum allium, aut cepe feriant. Illaq; per alembicum destillata, aquam uulneribus mundificandis, curandisq; utilissimã prestant, quæ & putridam carnem erodant.

Gauchblům.

D

Incognita ueteribus hæc hærba, tametsi passim in Germania nascens. Pedicularem sunt qui uelint, quod pedibus inimica, illos pellat, dum eius lixiuio caput lauatur.

Figure 5.4 *Gauchblům*. Brunfels, 1530. Illustration based on original watercolor by Hans Weiditz. Courtesy: Oak Spring Garden Foundation Library.

herbae nudae, preventing the reader from comparing the textual description of the plant with the illustration. The absence of a full textual description actually highlights the abstraction from watercolor to black line drawing to woodcut and draws attention to the limitations of printing in the early sixteenth century. In the introduction to *Contrafayt Kreüterbůch*, the German-language version of his herbal published in 1532, Brunfels discussed the illustrations, forty-six of which were newly cut for the new edition, while the remainder were reused from the 1530 Latin edition. He referred to the illustrations as "substitutes" for observation of living plants *in situ*, in gardens and herbariums. While the publisher Schott was keenly aware of the monetary value of the illustrations, Brunfels remained skeptical, despite their usefulness in clearing up uncertainties. For example, Brunfels could not find a counterpart in the ancients for the herb *Kuchenschell* (see Figure 5.5), also known as *Ruchschell or Sacketfraut*, and decided to discuss it only because the illustrations of other plants had been delayed. In order to keep things moving, he included it, so that writers who came after him could

corum TOMVS Primus. 217

A

B

Kuchenschell. Hacketkraut.

OTHO BRVNNFELSIVS.

CONSTITVERAMVS ab ipso statim operis nostri initio,
quicquid esset huiuscemodi herbarum incognitarum, et de qua=
rum nomenclaturis dubitaremus, ad libri calcem appendere, & eas tan=
tum sumere describendas, quæ fuissent plane uulgatissimæ, adeoq; & of
ficinis in usu: uerum longe secus accidit, & rei ipsius periculum nos edo=
cuit, interdum seruiendum esse scenæ τῷ καιρῷ λατρεύειν, quod dicitur. Nam
cum formarum deliniatores & sculptores, uehementer nos remoraren=
tur, ne interim ociose agerent & pręla, coacti sumus, quamlibet proxime
obuiam arripere. Statuimus igitur nudas herbas, quarum tantum nomi
na germanica nobis cognita sunt, prętera nihil. Nam latina neq; ab me=
dicis, neq; ab herbarijs rimari ualuimus (tantum abest, ut ex Dioscoride,
uel aliquo ueterum hanc quiuerimus demonstrare) magis adeo ut locum
supplerent, & occasionem prębérent doctioribus de ijs deliberandi, cp
t 3

Figure 5.5 *Kuchenschell*. Brunfels, 1530. Illustration based on original watercolor by Hans Weiditz. Courtesy: Oak Spring Garden Foundation Library.

have something to work with. That "something" was mainly the illustration, which is anything but feeble. The plant is depicted in great detail, with a hairy stem and leaves profusely swirling together, the overall effect being tight, bold, and textured. There are two flowers, one a bud similar in profile to a tulip, the other in full bloom with a profusion of stamens. Some of the leaves appear to be wilted or otherwise worse for wear, but the overall appearance is wild and uncultivated rather than pruned and tidy, which makes sense since the purpose of Brunfels' herbal was to describe simple herbs with medicinal properties. In fact, *Kuchenschell* was later known as the *Pasque* flower (*Anemone pulsatille*).[43] Čermáková and Černá (2018) refer to Brunfels' "pasque" flower,[44] even though there is no entry in his index for a plant under that name. However, "pasqueflower" was a common name for the genus *Pulsatilla*, which Bauhin described in 1623. All of this nomenclatural confusion could have been settled with accurate, realistic illustrations. It is helpful to have the whirling formation of sepals and leaves described, but Weiditz's illustration provides far more identifying information than either Brunfels or Bauhin.

The importance of Weiditz's illustrations was not lost on other, contemporary writers. Just a few years later, a competing publisher, Christian Egenolph, was sued by Johannes Schott, for reusing in his German-language *Kreuterbuch*, published in 1532, the same prints created by Weiditz.[45] Flechsig (2017) delineates the details surrounding the legal case, in which Egenolph argued that, in effect, the woodcuts could not be "copyrighted" because they depicted plants found in nature. Schott had made his claims to copyright clear when he issued the German-language version of Brunfels' herbal that same year. There, on the title page itself, above the imprint, Schott stated that the book was "not to be reprinted for five years, under penalty of five marks of full-weight gold."[46] In the Latin counterfeit *Herbarum Imagines Vivae* (Living Images of Herbs), in which Egenolff reused Weiditz's woodcuts, it is quite difficult to tell that they are Weiditz's illustrations because they have been so distorted, especially in terms of size and coloring. It is worth asking if anyone would have recognized them as Weiditz's were attention not brought to them via the lawsuit.

In 1542, Leonhard Fuchs (1501–66) produced his own herbal, *De Historia Stirpium* (An Inquiry into Plants). Not only did Fuchs identify *Gauchblům* as *Nasturtium agreste*, but he also planned to provide equally realistic illustrations, using three different artists: Albrecht Mayer, Heinrich Füllmaurer, and Veit Rudolf Speckle, who drew from nature, created line drawings, and engraved the woodblocks, respectively. Fuchs so appreciated their contributions to his text that he included on the last page a portrait of all three artists at work. Fuchs approached the illustrations with considerable forethought; he wanted the plants to be realistic but also ideal, which is better suited to identification. Beyond this, Fuchs' rationale has never been entirely clear, but Kusukawa (2012) draws a connection to his dispute with Sébastien de Monteaux, a French physician, regarding "accidents" as understood by Porphyry in his *Isagoge* to Aristotle's *Categories*. For Fuchs, the physical features of a plant were "native accidents" and thus inseparable from the substance of the plant. If so, the more specific, unique detail in an illustration, the better. In contrast to Brunfels, who recycled images of plants to depict others, especially in the third volume of the *Herbarum Vivae Eicones*, Fuchs may have been inspired by his feud with Monteaux to devise a methodology of plant illustration in which image exactly matches the plant.[47] Some of the plants were purposefully drawn as compound images, depicting them in various stages of their growth cycle, all in one image. This technique resulted in a hyper-realistic image, where an illustration of a flowering plant might also include withering leaves and other features not associated with spring and summer. This resolved the issue of space in a printed text, eliminating the need for multiple images of one plant, but reintroduced the ancient problem of identification in the wild. Whatever his motives, Fuchs understood the importance of illustrations, which is why Kusukawa traced the genesis of scientific botanical illustrations back to him, not Brunfels, criticizing Arber for having dismissed the importance of Fuchs' work in the history of plant taxonomy.[48] Arber regarded Fuchs as a mere regurgitator of Dioscorides, describing his writing style as "limited and imitative."[49] Given that the same criticism was leveled against Brunfels, historians of botany assume that the mere presence of realistic illustrations in a botanical text is not sufficient to warrant inclusion in the scientific field, even as an early practitioner whose work informed others who succeeded them. The illustrations must be aligned with the text and nomenclature and must be utilized with care and precision, which Brunfels failed to do. Historians of art, however, tend to view the sixteenth-century botanical illustrations with more grace. Egmond (2017) describes botanical printed illustrations of the time period as "visual manipulation," the result of which was "fairly unnatural representations that nonetheless look realistic."[50] This constitutes "scientific representation" characterized by a change in focus from the individual specimen to the essential, defining characteristics of the genus or species.[51] This definition aligns with Fuchs, who tried to show generic features rather than typical ones, but it also reveals a conundrum about precision. If the goal is to show the broad characteristics found in every typical specimen of a species, then the artist is constrained by the reference specimens to which they have access. If there is only one specimen, too much precision in the rendering is to be avoided, since imperfections such as disease or damage that are unique to that specimen may find their way into the illustration. This is the substance of some of the criticism leveled against Brunfels. At the same time, if one has a large number of specimens, some process must be devised by which the sum of their details can be systematically and accurately distilled into one illustration that manages to depict the defining characteristics of them all while still maintaining a high level of precision.

An indication of the precision with which Fuchs approached illustrations is that he intended his book to be colored in after printing. In anticipation of this, he made a point of describing multiple varieties of the same plant, making note of their coloration, with the intention of a colorist adding in the correct pigments post-production. This process is not clear in extant uncolored copies, but is evident in surviving colored copies. For example, a plum tree is colored with rosy-red on the right, blue in the middle, and yellow on the left, to represent the three types of plum. Similarly, *maize*, which Fuchs incorrectly calls Turkish, rather than American,[52] is depicted with four different illustrations. In some copies these are colored according to the textual description, which indicates the four common colors of corn: yellow, purple-red, red, and white, in that order. A single rose illustration is painted with red on the right and white on the left, and so on. There were staff colorists in Antwerp at the printing establishment of Christopher Plantin (*c.* 1520–89), the French publisher and printer, so it is not unlikely that Fuchs or his publisher had access to similar staff artists. However, the skill level is generally inferior to that of some extant colored copies of Brunfels. For example, in some copies of Fuchs, the color palette is limited, using only a couple of shades of green for the stalk and leaves, and shades of red or pink for the flowers. In such cases, the coloring is not scientific, nor does it even count as realism in art, but appears to serve the basic function of adding color to the images so that they are not just black and white. In other copies, flowers and highlights are depicted with a haphazardly applied whitewash splash, such as *Silphium* (*Laserpitium*), where in one copy someone simply smeared white pigment and a light teal over the flowers with no attempt at precision.[53] There is also a lack of concord between copies, with some showing white flowers for *Nasturtium agreste* and others rendering them in pink.[54] Similarly, *Consolida media* has blue flowers with leaves that are pink on top in some copies, while others depict flowers that are very dark, almost midnight blue, and ruddy on the top.[55] The technology of printing full-color images had not yet developed at this point and without staff colorists, Fuchs would have known that once a copy sold, he was relinquishing control to the buyer. The variations in colored-in copies of Fuchs demonstrate a further level of abstraction from physical, living specimen to post-production coloring processes. No matter how carefully Fuchs and his artists planned out their illustrations, a buyer could, with a few random brush strokes, render the image unscientific. Fuchs' illustrations were also sized down for later quarto editions, which depicted four images per page, without much loss of detail. They were also used as references for illustrations for Rembert Dodoens (1517–85) and adapted by David Kandel for Hieronymus Bock's (*c.* 1497–1554) *Kreüterbuch*, meaning that abstraction and distortion could also originate in the printers and publishers themselves.

The difference between the drawings of Fuchs' artists Mayer and Füllmaurer and those of Weiditz can be seen by comparison with *Gauchblům*, which Fuchs classified as *Nasturtium agreste* (see Figure 5.6). In Fuchs, the lines are much

Figure 5.6 *Nasturtium agreste* (Gauchblům). Fuchs, 1542. Illustration and woodcut by Albrecht Mayer, Heinrich Füllmaurer, and Veit Rudolf Speckle. Courtesy: Oak Spring Garden Foundation Library.

finer and more delicate. There is a sparseness to the image and an economy of expression in which it feels like the bare minimum information has been provided. There is no need to depict insect damage, blight, frost, and other imperfections. If the goal is to render the plant discoverable to practitioners, it is sufficient to provide exactly enough detail to make that a possibility. This is especially true when one is attempting to identify the taxonomic rank of a plant that a fellow practitioner had earlier failed to identify. There is a similar situation with Fuchs' illustration of *Consolida media* (see Figure 5.7), about which Brunfels had been unsure of the proper nomenclature. In Brunfels, the illustration occupies only the top left quarter of the page and is compressed in such a way that the details are muddied, while Fuchs' illustration is afforded a full page. The details in each illustration are much the same, both depicting the tangled roots, the tall, curved stalk, the placement of the leaves, and the flower stalk, which is

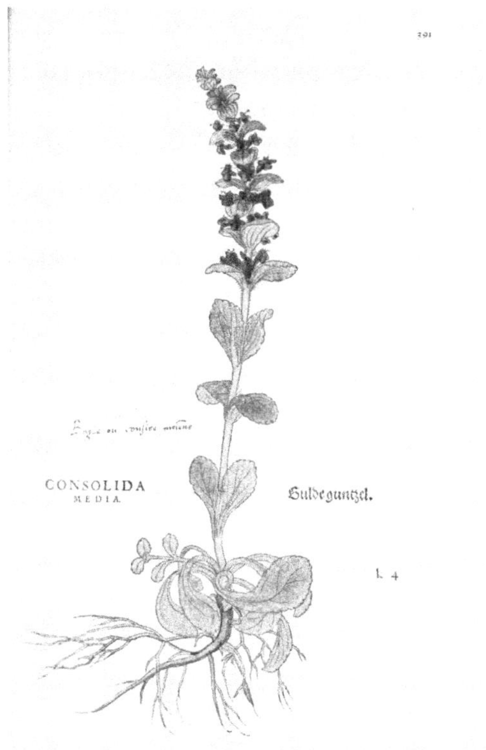

Figure 5.7 *Consolida media*. Fuchs, 1542. Illustration and woodcut by Albrecht Mayer, Heinrich Füllmaurer, and Veit Rudolf Speckle. Courtesy: Oak Spring Garden Foundation Library.

profuse with flowers. The shape of the leaves, however, is different in each illustration. In Brunfels they are pinnate, while in Fuchs they are rounded and oblongate. In Brunfels, the stalk, leaves, and even some of the flowers are depicted as hispid (hairy), indicating an overall hirsute herb, a detail that is missing in Fuchs. Apart from this, the illustrations are quite similar, but, where Brunfels simply listed its medicinal benefits, Fuchs included a textual description of its physical features and habitat.

Despite the differences between Fuchs' and Brunfel's illustrations, they share enough commonalities that their significance for scientific study is better understood by comparison with other, roughly contemporary texts. Just a few years before Brunfels, Peter Treveris (active 1525–32) had produced in 1526 his English language *The Grete Herball* in London. Like Brunfels and Fuchs, Treveris provided an index of names with Latin nomenclature matched to their English counterparts. Unlike them, his woodcut illustrations are simplistic in a style typical of the early printing era and are not detailed enough to guarantee discovery in the wild. For the herb labelled *Nasturcium agresti wylde cresses*, Treveris not only did not bother with realism, but the plant depicted is likely not even a species of *Nasturtium* (see Figure 5.8).[56] The herb shown has one straight stalk with four branches, two of roughly equal length on each side, but with no consistency in the number or dispersion of the leaves, which are roughly pinnate in shape. The lack of detail is not the biggest issue, however, as economy of expression can be used efficiently. In Treveris, the plant depicted looks nothing like either *Gauchblům* in Brunfels or *Nasturtium agreste* in Fuchs, but is instead emblematic of the real herb, a standard, late Renaissance woodcut that vaguely approximates the plant in question. In 1561, twenty years after Fuchs, Antoine du Pinet's (*c.* 1510–84) *Historia Plantarum* also depicts a *Nasturtium aquaticum* (see Figure 5.9). Neither text nor image in this tiny book are particularly original. In many entries, du Pinet cited Fuchs and Pietro Andrea Mattioli (1501–*c.* 1578), the naturalist most famous for his commentary on Dioscorides. Du Pinet's woodcuts are very similar in style to those of Mattioli, and although small, the illustrations manage to convey a significant amount of information. In particular, the growth pattern, overall shape, and the form of the leaves of *Nasturtium aquaticum* are clearly shown and are similar enough to the illustrations of *Gauchblům* in both Brunfels and Fuchs to indicate that this is the same genus, if a bushier species. No flowers are depicted in du Pinet, likely for a combination of reasons, including the relatively small size of the book. Du Pinet needed only to depict the parts of the plant most pertinent to the textual descriptions, which are quite limited in scope. They generally involve a list of the nomenclature in Latin,

Figure 5.8 *Nasturcium agresti wylde cresses*. Peter Treveris, 1526. Courtesy: Oak Spring Garden Foundation Library.

Greek, Italian, German, Spanish, and French; the habitat; and a few simple "qualities," such as taste and categorization in the order of heating and drying plants.

The sandwiching of Brunfels and Fuchs by Treveris and du Pinet highlights their liminality. The similarity of Treveris' and du Pinet's illustrations to the medieval and Renaissance herbals is to be expected, since the turn to realism was fairly recent and, at least initially, constrained to Germany. The expense of producing original, realistic woodcuts was not always justifiable to publishers and printers, who continued to use and reuse simpler illustrations and woodblocks that could be adapted for more than one text. The Flemish printer Jan Moretus (1543–1610) later bought, commissioned, and lent out to other publishers thousands of botanical woodblocks, with which he printed two editions of Rembert Dodoens. Some were lent to a London publisher, for only one edition of their book, although they reused them for a second edition before returning them to Antwerp. Good woodblocks were valuable and could be used

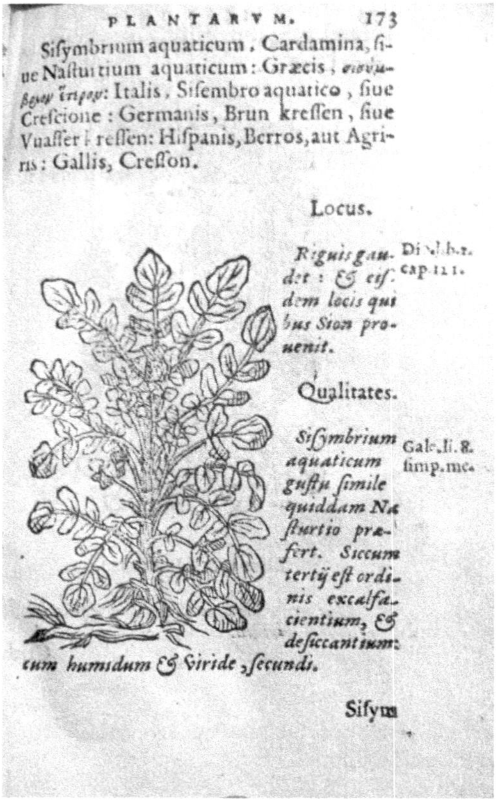

PLANTARVM. 173

Siſymbrium aquaticum, Cardamina, ſiue Naſturtium aquaticum: Græcis, σισύμβριον ἕτερον: Italis, Siſembro aquatico, ſiue Creſcione: Germanis, Brun kreſſen, ſiue Vuaſſer kreſſen: Hiſpanis, Berros, aut Agrias: Gallis, Creſſon.

Locus.

Riguis gaudet: & eiſdem locis quibus Sion prouenit. Di. li. 2. cap. 121.

Qualitates.

Siſymbrium aquaticum guſtu ſimile quiddam Naſturtio præfert. Siccum tertij eſt ordinis excalfacientium, & deſiccantium: cum humidum & viride, ſecundi. Gal. li. 8. ſimp. me.

Siſym

Figure 5.9 *Nasturtium aquaticum*. Antoine du Pinet, 1561. Courtesy: Oak Spring Garden Foundation Library.

in different countries, regardless of language. Chrisopher Plantin, for whom Moretus had been an apprentice, also continued using his stock of botanical woodblocks despite the advent of newer techniques. By owning his woodblocks, Plantin was able to control the market, quite apart from issues of intellectual property. This helps with understanding Fuchs, whose more idealized images may have been at least partially in response to the Schott versus Egenolph lawsuit. Publishers generally commissioned the artists and owned the results of their work. Schott's lawsuit was anomalous because it was industry standard to "steal" illustrations from one another. He may have been naive to think he could win such a suit based on claims of intellectual property and may have had better success arguing for lost profits. Either way, by commissioning such idiosyncratic illustrations, which were so closely tied to the specific plants described, and were produced by three clearly named artists, Fuchs and his publisher ensured that their woodblocks had less market value for others.

Just as printed herbals continued in the sixteenth century to use and reuse woodblocks of an earlier style, so too did manuscript herbals persist. Even when an author intended to print their work, it was common to assemble the herbs in manuscript beforehand. Between 1555 and 1565, Conrad Gessner (1516–65) compiled in manuscript form under the title *Historia Plantarum* a more traditional, but idiosyncratic, herbal encyclopedia.[57] In keeping with the herbal tradition, where authorship was loosely defined and texts were often added to over time, each plant in Gessner has copious and haphazard, handwritten notes. Daston (2004) argues that this technique is a crucial part of the process of scientific observation, claiming that it "binds together the practices of observing and reading."[58] Gessner's illustrations were made by various individuals, sometimes Gessner himself, although in some cases the artist is unclear.[59] Gessner did not always sign individual leaves or make a clear attribution to someone else. The illustrations include many of the usual botanical suspects, such as absinthe (*Artemisia absinthium* L., fol. 185r) and opium (*Papaver somniferum* L., fol. 274v), but the herbal diverges from tradition in a few significant ways, displaying both Gessner's scientific interests and the turn to the so-called "New World," whose specimens were now finding their way to Europe. For instance, fol. 13r depicts *Nicotiana rustica* L., and in addition to the full tobacco plant, Gessner has also shown the flowers, seed pods, and leaves close up, an early example of the zooming techniques[60] that ramped up in botanical illustrations from about 1580 on.[61] There are other New World plants, such as the tomato (*Lycopersicon esculentum* L). on fol. 37v and fol. 42r, the former showing a close up of the leaves and fruit, while the latter shows the entire plant with a few small fruits in different stages of ripeness.[62] Kusukawa (2015) notes that Gessner did not initially know what the tomato plant was. He had encountered a specimen in 1654 but did not recognize it, so he continued to draw and make notes on it until he learned its origins and common name.[63] This technique is also seen in the Codex Kentmanus, a manuscript record consisting of watercolors and observations of the plants in the botanical garden in Padua, compiled by German physician Johannes Kentmann (1518–77). Kusukuwa (2009) argues that Kentmann steadily synthesized information about the plants from apothecaries, the written herbal tradition, and other medics to produce accounts of the plants that were less specific but more useful for physicians. Gessner borrowed Kentmann's manuscript, copied it, and returned it in March 1555, proof that Kentmann had finished it by 1554 at the latest.[64] By the mid-sixteenth century, there was a developing caution in both textual descriptions of plants and their illustrations. Both Kentmann and Gessner were at leisure to compile as much information as possible about plants, to cross-reference their own observations with others to ensure accuracy in terms of physical characteristics, nomenclature, and classification. Kentmann had access to living specimens at the botanical garden, while Gessner, in some cases at least, had to be satisfied with dried specimens or even simply textual accounts of natural phenomena. Gessner therefore synthesized and compiled these sources, adding them to his herbal piecemeal.

A particularly striking feature of Gessner's illustrations, especially since they were not limited by developing printing technique, is that some are only partially colored in. For example, on fol. 26r, the illustration of *Carduus nutans* L. is colored in on the top half but not the bottom half, which is a mere sketch outline. On fol. 36v, the illustration of *Thalictrum minus* L. is colored in on the bottom and top while the middle is not.[65] In other cases, there might be two images on one folio, one in color, with a black-and-white "shadow" line drawing in the background, for example fol. 455v, depicting *Brassica oleracea* L. Some folios have nothing but sketches of one or more plants while others have an outline that was never colored in, such as on fol. 57v, which shows *Pulmonaria obscura* Dumort. On fol. 58r, the same plant is colored in and is signed by Josef Murer, whereas on fol. 57v the artist is unknown. Meanwhile, fol. 57r also depicts this plant in full bloom, although this illustration has Gessner listed as the artist. Of the three images of the same species, the first is done by Gessner and shows the plant in full bloom, possibly even close to senescence, based on the shape. The second, uncolored and unsigned image has no flowers and appears to show it in a stage of dormancy, or just before bloom. The third image by Josef Murer shows it in the early stages of flowering, with an overall shape that is more compact than on fol. 57r, suggesting it has time before it goes to seed. Regardless of how the posthumously printed version of this herbal would eventually depict this plant, and regardless of how that illustration may have been colored in post-printing, Gessner ensured that he had documented the plant as thoroughly as possible. Generally speaking, he focused on the morphological characteristics of the plants and was primarily interested in their identification and medicinal parts, which informed his decision to show more than one stage of growth. The herbal provides insight into and evidence of Gessner's techniques, references, and ways of thinking about and knowing plants. For instance, there are a few leaves into which are pasted cut-outs of printed woodblock illustrations from other texts, such as on fol. 349r, which has two separate reference images. In addition, there are a couple of nature prints on fols. 326r and 327r, suggesting that he recognized the value in this form of "artless" preservation. Specimens dry out and lose much of their vivacity, but by inking a specimen and printing it, it is possible to preserve and essentially illustrate many of its features, especially finer details like leaf venation.[66] Florike Egmond has worked extensively on the non-printed, original drawings of plants and animals that proliferated in the sixteenth and seventeenth centuries. She argues that there is more to these nature drawings and paintings than "epistemology" and that, paraphrasing Pächt, "images should ultimately be worked out visually, with an understanding of the cultural codes or visual languages of their time."[67] Like Gessner, the Swiss physician and botanist Felix Platter (1536–1614) cut and pasted Weiditz's watercolors into his own herbarium, in close proximity to dried specimens of the same plants, and also cut out printed illustrations from the proofs of botanical texts in the same way.[68] Recalling Brunfels and Fuchs, on fol. 364r Gessner illustrated *Cardamine pratensis* L., the Linnaean reclassification of *Nasturtium agreste* (Gauchblům). He did not use either of these names, however, instead calling it *Wiesen-Schaumkraut*, a common name for "Cuckoo Plant." Gessner was the illustrator for the fruits, on which he focused far more than he did the full plant, which fades in the background compared to the oversized depictions of seeds.[69] Gessner would have known that the plant had already been depicted by Brunfels and Fuchs and for this reason, he chose to pay closer attention to certain identifying parts that were not attended to or overtly depicted in their herbals.

In contrast to Gessner's illustrations, Ulisse Aldrovandi (1522–1605) focused primarily on physical specimens for his *Erbario* (Herbarium), begun around 1551. He worked with various artists, including Jacopo Ligozzi, Giovanni Neri, and Cornelio Schwindt, to illustrate the specimens collected on the many expeditions he funded from roughly 1551 to 1554. Aldrovandi was concerned with amassing a large collection, not with the systematic classification of the plants, and did not always bother to verify the plant names or map them to those in other traditions and languages. Aldrovandi was influenced by his colleague Luca Ghini (1490–1556), who is credited with the first such herbarium (*hortus siccus*, or dried garden) and the first formal botanical garden in Europe, which became the Orto Botanico di Pisa. Ghini taught both Gherardo Cibo (1512–1600), who compiled a herbarium using the same preservation techniques,[70] and Pietro Andrea Mattioli, whom he assisted in his travels around the Mediterranean in order to identify the plants described in Dioscorides' *De Materia Medica*. It is no surprise, then, that Aldrovandi was aware of and influenced by the "discoveries" being made in the Americas. For example, there are three leaves of dried specimens of *Nicotiana tabacum* L., two with just the leaves in different arrangements and one with leaves and flowers.[71] Like Gessner, Aldrovandi saw fit to collect and describe American plants, including several species of citrus[72] and even American corn (*Zea mays* L.).[73] The illustration of the latter is based on a dried specimen, which is clear from the gathering of the male flowers in an inflorescence. Because the fruit, the ear of the corn, is hardly visible, its most recognizable feature is missing. This is unfortunate, since "granoturco," which Mattioli had argued should be called Indian because of its origins in the West Indies, had begun to be cultivated in Tuscany by 1594 and was therefore known well enough to warrant an accurate illustration. One particular case highlights the implications of relying on hearsay or dried specimens to produce illustrations. The plant listed as *Persea americana* (Pers.) Mill. is in fact not clearly identified. Based solely on the illustration, it is debatable whether it is one of four different plants: avocado (*Persea americana* or *Persea gratissima*),[74] Mamey sapote (*Lucuma mammosa*), papaya (*Carica papaya* L.), or Mamey (*Mammea americana* L.). Aldrovandi's artists based the image on the one in Pietro Antonio Michiel's (1510–76) herbal as well as on a specimen sent to him in 1568. All of these fruits look similar enough to avocado to render the illustration problematic for identification. It is the specimen, even though it is just leaves, that provides the necessary verification. This interplay between specimen, text, and illustration shows that identification, knowability, and intelligibility are dependent on the collaboration of all three. Throughout the sixteenth century, there was a developing recognition amongst European botanists that specimens and realistic illustrations were indispensable for identifying, classifying, and ultimately sharing knowledge about plants.

Textual descriptions were also developing into a scientific genre, in tandem with illustrations. In 1576, Carolus Clusius (1526–1609), whose Latin descriptions show a turn toward a more scientific approach to botanical texts,[75] produced his *Rariorum aliquot stirpium*. The illustration for *Siliqua silvestris* (see Figure 5.10) also demonstrates a style of illustration that more closely resembles the conventions of modern botany. Rather than a full-length portrait of the tree, the illustration zooms in on a branch, showing the leaves, the flowers in a few different stages from bud to full bloom, and the primary feature of the tree for which it is named, its seed pods, which resemble peapods. The leaves are heart-shaped, similar to those of *Cercis canadensis* (Eastern redbud), and are depicted with an attention to detail, showing their veins and even managing to include the axillary buds just above the intersection of the petiole and the stem. The inclusion of such details is not happenstance; it is the effect of a turn toward closer

Figure 5.10 *Siliqua silvestris*. Carolus Clusius, 1576. Courtesy: Oak Spring Garden Foundation Library.

inspection of plant parts and the recognition that depicting certain parts is crucial for conveying taxonomic information to one's readers. Clusius was not depicting the plant for herbalists and physicians, but for his fellow botanists. Every feature of the illustration therefore provides identifying information. The accompanying text also displays many of the features of formal botanical Latin, including a shorter diagnosis followed by a much lengthier description, ablatives in paratactic constructions, and more specialized botanical vocabulary. The trend toward illustrating plant parts separately had already appeared by 1563, when Rembert Dodoens published his *Cruÿde boeck* in Antwerp with small woodcut illustrations very similar to those of Weiditz. In 1578, the first English edition was published under the title *A Niewe Herball*, having first been translated into French by Clusius himself. The illustrations run the gamut from full-length images of trees for which the overall growth pattern is relevant information, for example, *Cypress*,[76] to cross-sections containing portions of the trunk with roots, a branch, leaves, flowers, and fruit. In the case of *Oxycedrus* (Cedar), the trunk, bark, and roots are all secondary to the branch, which is shown with an intense profusion of cedar needles and berries that are not in proportion to the trunk (see Figure 5.11). He began the entry by saying that the fruit of the smaller species is "of two sortes: the one with sharpe prickley leaves like Juniper; the other are not prickley at all."[77] The textual description matches this focus on the branch and needles, which he called leaves. Later, he said that the "branches be clad and garnished with many smal little leaves, thicke, short, and sweete smelling." It is clear from the information provided by Dodoens about the kinds of cedar that an image of the entire tree would not be as helpful for identification as the illustration depicting a close up of the fruit and needles. Most illustrated botanical texts provided images of all of the plants described, usually either on

764 The syxth Booke of

This gumme tempered with Oyle of Roses, helpeth the riftes, cones, or chappinges of the handes and feete.

Of Cedar tree. Chap.lxxxiij.

The Kindes.

There be two sortes of Cedar, great & small. The smal fruite also is of two sortes: the one with sharpe prickley leaues like Juniper: the other are not prickley at all.

The Description.

Oxycedrus.

1 The great Cedar wareth very stowte & tall, high, great & thicke, yea greater, & higher then the figge tree. The barke euen from the foote of the stem unto the first branches is rough, and from thence foorth euen vp to the toppe, is very smoth & playne, of a darke blew colour, out of which there droppeth white Rosen of his owne kind, which is moyst and odoriferant or sweete smelling, the which by the heate of Sonne becommeth dry and harde. His limmes and branches be long, and stretched out into length & breadth and parted into many other small branches, standing directly or right one against another, lyke as in the Firre tree. The sayde branches be clad and garnished with many smal litle leaues, thicke, short, and sweete smelling like ŷ leaues of Larix, or Larche tree. The fruit is like that of ŷ Firre tree, sauing that it is greater, thicker, & harder, & the tree groweth straight vpright like the Firre tree, as the paineful & diligent Peter Belon hath written. From the tronke or stemme of the Cedar tree, there commeth foorth a certayne cleare liquor, which the olde writers called Cedria.

2 The first kind of ŷ smaller Cedar, is much like to Juniper: but most cõmonly it is somwhat smaller. The stem is croked or writhed, & couered with a rough barke. The fruit is round beries, like Juniper beries, but somwhat greater, in colour at ŷ first greene, then yellow, & at last reddish, of an indifferent good tast.

3 The second kind of smal Cedar groweth not high, but remayneth alwayes smal and lowe lyke the other. The leaues of this kind are not prickley, but somwhat round & mossie at the endes, almost lyke to the leaues of Tamariske and Sauin. The fruite of this kinde also is rounde beries greene at the first, afterwarde yellow, and at last reddishe, in taste bitter.

The Place.

The great Cedar groweth in Africa and Syria, and as Vitruuius reporteth

Figure 5.11 *Oxycedrus* (Cedar). Rembert Dodoens, 1578. Courtesy: Oak Spring Garden Foundation Library.

the same page or in close proximity to the text. Joachim Camerarius (1534–98), on the other hand, published in 1588 his *Hortus Medicus et Philosophicus*, which is unique because the first part includes textual descriptions while the images (*icones*) are in a separate section with its own title page, *Icones Accurate.* Camerarius did not bother to provide images for all of the plants he described, so this is a much shorter section. There is also a separate catalog of trees that grow in Hercynia. Like du Pinet, the illustrations look quite similar to those of Mattioli. Camerarius only provided some illustrations, so it is worth noting that he depicted *Nasturtium indicum* twice, on plates XXXI and XXXII, showing first the entire plant and then the flowers with seeds in separate illustrations (see Figures 5.12 and 5.13). In this case, the most efficacious method of ensuring accuracy in the depiction of the plant was to show the full view and a close up of its distinguishing characteristics. *Nasturtium* and other similar genera had been plaguing herbalists and proto-botanists since at least Fuchs and a sixteenth-century herbal was hardly complete without at least one species.

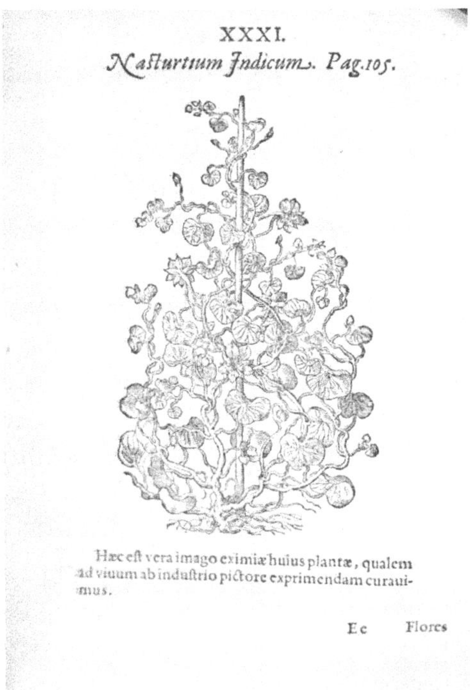

Figure 5.12 *Nasturtium Indicum.* Joachim Camerarius, 1588. Entire plant. Courtesy: Oak Spring Garden Foundation Library.

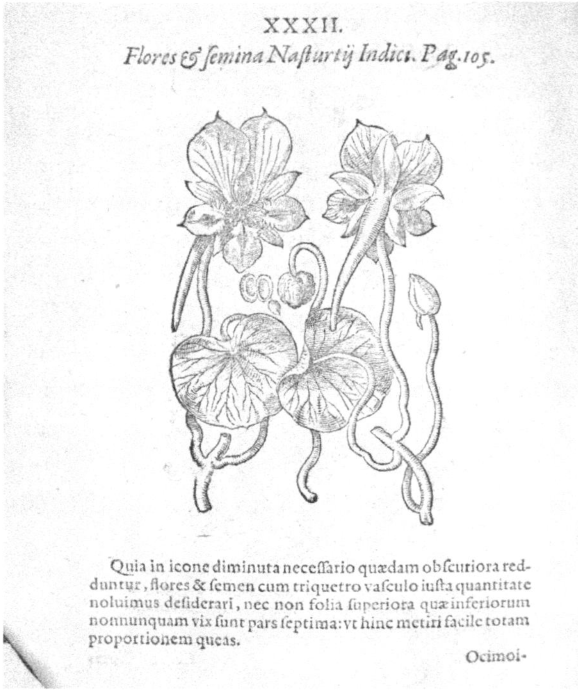

Figure 5.13 Flowers and seeds of *Nasturtium Indicum*. Joachim Camerarius, 1588. Courtesy: Oak Spring Garden Foundation Library.

The year 1592 saw the publication of both Fabio Colonna's (1567–1640) *Phytobasanos* and Prospero Alpini's (1553–1617) *De Plantis Aegypti Liber*. Colonna's full-page etchings are situated inside a pre-printed woodcut decorative border, which is the same for every illustration. The etchings show exceptional precision, evidence of progress in the technique of printing illustrations. Many of the plants are from the Greek tradition and bear names such as *Isopyrum Dioscoridis* or *Papaver Heracleum Theophrasti*.[78] From that same tradition is the plant *Phyteuma Dioscoridis*, about which Pliny had stated that he considered it a waste of time to describe it, since it was used only "for the art of love."[79] Colonna's illustration is meticulous, showing the entire plant, including the roots, flowers that have gone to seed, upper and lower leaves with distinct and diverse shapes, and even parts of the flowers dissected out (see Figure 5.14). The image is an early printed example of a botanical convention, in which the identifying features of the flowers and fruit are shown in order to help the reader envisage the accompanying textual descriptions. The technique was not uncommon in manuscripts such as Gessner's, but it had yet to become standard in printed texts.

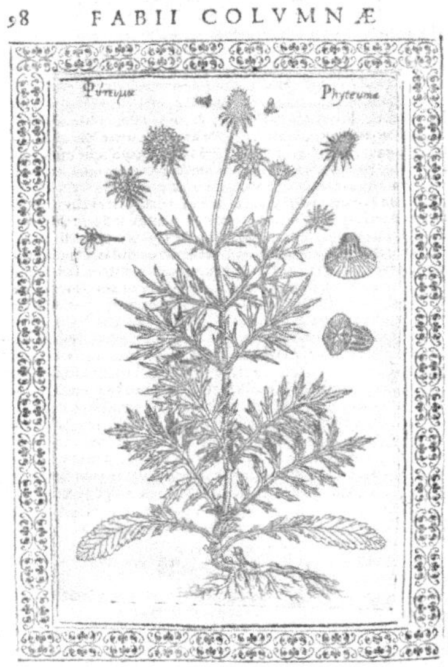

Figure 5.14 *Phyteuma*. Fabio Colonna, 1592. Courtesy: Oak Spring Garden Foundation Library.

Colonna cited Dioscorides, Pliny, and even Aristophanes, but provided his own description of the physical features of the plant, indicating a close relationship with the artist and etcher, who would have been instructed which parts of the plant to highlight in this way. Alpini's text, meanwhile, shows how botanists were beginning to look abroad in order to find plants to describe. The flora of Western and Northern Europe had been cataloged, mapped, identified, and described, and botanists were keen to do the same to plants unknown to them. Alpini even provided in his preface an *apologia* for his book entitled "Why it is that [a book] on plants growing in places in Egypt must be made."[80] Many of the plants are not native to Europe and therefore had not been described in Latin or mapped to Greek or Latin nomenclature from antiquity. They could certainly have been known in antiquity under other names, and even perhaps described in ancient herbals, but they are not found in the few that are extant, despite Egypt having been under both Greek and Latin rule. In the Roman botanical tradition, Egyptian plants, or plants with names indicative of origins in Egypt, were not uncommon. The fact that Alpini provided only the Romanized Arabic names indicates that they were not known to the Greeks and Romans. Taking Colonna's dissection method of illustration to another level, Alpini included full-page, separate illustrations of certain parts of the plants, for example, the fruit of the Baobab tree, which is shown in cross-section so that the rind and seeds are visible. For *Agiahalid* (fol. 16v), he depicted the tree in full, without the roots, which are unimportant for identification (see Figure 5.15). Five twisted branches are shown growing almost straight up, a somewhat unusual growth pattern, though not uncommon in orchards or in other areas where trees compete for space and light and are forced to grow in an upright fashion. The top portion of each branch is bare, so that they appear almost like oversized thorns. The leaves are shown in pairs, each with what looks like a thorn projecting above it, while the textual description states that the tree has thorns (*spinis*), only a few branches, and sparse leaves. The proportions are off, with the branches, leaves, and flowers far larger than in real life. There is, nevertheless, sufficient information encoded in the image to assist in identification in the wild. Ironically, the specimen from which Alpini worked is cultivated, as he stated that it was brought from Ethiopia

Figure 5.15 *Agiahalid*. Prospero Alpini, 1592. Courtesy: Oak Spring Garden Foundation Library.

to the arboretum (*viridarium*) of the Eunuch Turca, whose collection he consulted for several other plants in the book. The popularity of such collections, whether living, in the form of botanical gardens, hothouses, and arboretums, or as dried specimens is directly correlated with the increase in world travel. Some botanists did their own traveling, collecting specimens and sketching along the way, while others simply received specimens. These were usually dried or in some way preserved, although much later they began to arrive still living, when Wardian cases made this a possibility for overseas transport. In the absence of living specimens, botanists relied on the written notes and descriptions of the collector in order to supplement them. In this case, Alpini had the benefit of being able to observe a living specimen *in situ*.

In a similar vein, in the 1596 English translation of Nicolás Monardes' (1493–1588) herbal, *Joyfull Newes out of the New-found Worlde*, he discussed the flora and fauna of the Americas. Like Colonna, Monardes provided separate illustrations of fruits, flowers, seeds, etc. In the entry called "Of the flower of Mechoacan" (see Figure 5.16), a plant named *Mechoacan* after its place of origin is described.[81] It had already been discovered in "Newe Spayne," but in a recent voyage to Nicaragua and Quito, Ecuador, another species was found, specimens of which several individuals brought back. Because this new variety was used as an emetic and was described as more potent than the others, Monardes showed the flower, which is both the part used for this purpose and a sure way to distinguish this species from others. The flower is shown with seven long, rather limp leaves (possibly due to the difficulty of keeping a specimen both alive and healthy on a sea voyage) and an ovary/ovule with no stamens. In his description of the flower, he said that the leaves grow from a "blister" that resembles a nut and is divided into two parts by a vein, which is an accurate description of the image. Monardes also noted that explorers had brought back the fruit, the flower "as whole as it is in the plante," the leaves, and "bowes." This is unfortunate, since Monardes claimed in the prior paragraph that the roots are eaten along with other parts of the plant. However, he did note that the roots simply tasted nice, whereas the other parts had medicinal properties.[82]

In a twist on the turn to smaller woodblock illustrations to save space and money, Jacques and Paul Contant, father and son apothecaries, took efficiency to a new level in their *Les Oeuvres* of 1628.[83] The book consists of five parts, the first being a commentary on Dioscorides, the second a long poem called the *Second Eden*, and the third a treasury of the benefits of herbs. The fourth deals with etymology, which is useful for those who may not know Latin and Greek well, recalling Pliny's insistence on descriptive plant names. The audience for the text

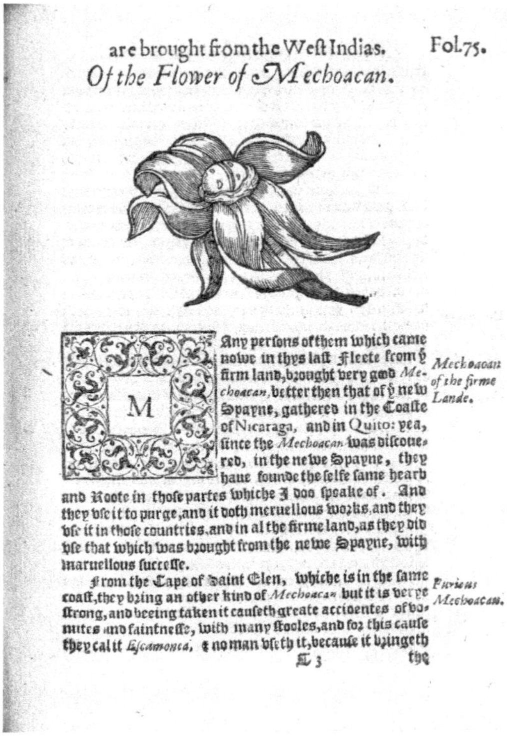

are brought from the West Indias. Fol.75.

Of the Flower of Mechoacan.

Many persons of them which came nowe in thys last Fleete from ỹ firm land, brought very good *Mechoacan*, better then that of ỹ newe Spayne, gathered in the Coaste of Nicaraga, and in Quito: yea, since the *Mechoacan* was discouered, in the newe Spayne, they haue founde the selfe same hearb and Roote in those partes whiche I doo speake of. And they vse it to purge, and it doth meruellous works, and they vse it in those countries, and in al the firme land, as they did vse that which was brought from the newe Spayne, with maruellous successe.

Mechoacan of the firme Lande.

From the Cape of Saint Elen, whiche is in the same coast, they bring an other kind of *Mechoacan* but it is verye strong, and beeing taken it causeth greate accidentes of vomites and faintnesse, with many stooles, and for this cause they cal it *Escamonea*, & no man vseth it, because it bringeth the

Furious Mechoacan.

L 3

Figure 5.16 *Mechoacan*. Nicolás Monardes, 1596. The medicinal part of the flower is depicted. Courtesy: Oak Spring Garden Foundation Library.

consists of fellow apothecaries for whom it is useful to know what the plant names mean. Unlike other herbals, rather than individual illustrations on the same page or the opposite page as the textual description, the authors provided a full-page, untitled visual index on fol. 8r, consisting of ninety illustrations in a 9 × 10 grid, each image measuring about 22 × 27 millimeters (see Figure 5.17). The images are too small to render much detail, but are sufficient to give an overall sense of a plant's likeness. From a printing standpoint, it would have been quite costly to have full- or half-page illustrations. By consolidating all the plants discussed in their commentary on Dioscorides onto a single page, they cut down the costs of production. At the very end, as part of the fifth part of the text, *Le Jardin et Cabinet Poëtique*, on fol. 60r, is a plate depicting a vase of sorts chock full of plants. This takes up the entire page and the plants are also numbered, serving as a fanciful index mapping the images to their textual references in the poem. It is tempting to dismiss the Contants' text as mere whimsy and the illustrations as entirely impractical, but their efficiency and efficacy in contrast

Figure 5.17 Untitled visual index. Jacques and Paul Contant, 1628. Courtesy: Oak Spring Garden Foundation Library.

to contemporaneous botanical works serves to highlight the inherent limitations of botanical illustrations. If the cost and effort of creating and producing large-scale, detailed illustrations that assist in identification are out of reach, can one substitute with others that are "good enough"?

A few years later in 1631, Antonio Donati produced his Venetian-focused *Trattato de semplici*. Though not strictly a herbal, Donati's text nevertheless followed the trends in descriptions of "unknown" plants, while still retaining the herbal tradition's focus on medicinal plants. In this case, Donati focused on "simples," individual plants that do not need to be added to complex compounds and can generally be grown in kitchen gardens, or are at least common enough to be easy to find in the wild. Some of the descriptions therefore also include recipes for how to use them. On the leaves with illustrations, it is clear that the titles were printed originally with the text, while the rest of the page was left blank in anticipation of the illustrations. These were added after printing, so that in some cases, such as *Sideritis Eraclea*, the name is partially obscured by the illustration.[84] For the plant *Artemisia marina* (*Absinthium Marinum*), the illustration is both detailed and artistically impressive (see Figure 5.18).[85] The plant's bare roots are shown, together with the stalks, leaves, and flowers, which have a unique shrub growth pattern in which the leaves become increasingly minute the higher they are on both stalk and plant overall, creating a narrowing effect. The individual flowers are not the focus, but rather the overall shape of the plant, which grows into a woody, upright shrub, and the narrow, pinnate leaves, by which the plant is often recognized. The techniques of printing had developed sufficiently by this point that the artist was able to render shadows and depth, along with the texture of the stalk and leaves. Species of *absinthium*, genus *Artemisia*, are often identified by reference to their leaf color and shape first and foremost, with color and scent as close seconds, and taste a cautious third, given the toxicity that led to the

Figure 5.18 *Artemisia marina.* Antonio Donati, 1631. Courtesy: Oak Spring Garden Foundation Library.

development of absinthe. In Donati, despite the limitations of the black-and-white illustration, a reader would have little trouble identifying the plant.

This is in stark contrast to John Gerard's (1545–1612) *The Herball, or Generall Historie of Plantes*, originally published in 1597 and revised in 1632 and 1636. Both revisions, by Thomas Johnson, used the famous woodblocks of Christophe Plantin. In general, each illustration takes up about a quarter of the page and is interspersed with other illustrations and text. While they retain the detail of the original woodblocks, their compression in Gerard's pages means that they in some cases have a smudged look, especially where any attempt has been made to render a plant's texture. One illustration on pg. 1190 is of particular interest, *Coronopus ex codice Caesarea* (see Figure 5.19). According to the textual description on the following page, this image was produced by Dodoens and is based on the illustrations in "an old Manuscript in the Emperor's Library," an oblique reference to the Julia Anicia Codex of Dioscorides. Gerard frankly critiqued the accuracy of the depiction, complaining that the plant's five-leaf gatherings were incorrectly depicted, but that the overall effect is that the "parts are well exprest" and that any deficiencies could be attributed to the artistic style of the time period. He therefore provided a footnote with a description of the plant's growth patterns, as they are not shown in the image. In particular, the plant's shoots, which can grow as long as a cubit, are not usually upright, but rather creep along the ground. In addition, the leaves form groups of three not five, hence the names *Trefoile* and *Trifolium*. The flowers, which are not the focus of the illustration, grow in clusters at the apex of the stems and resemble those of the Burgundy species of *Trefoile* in both color and scent and in the shape of the *coddes* (seedpods), which Gerard described as narrow and crooked, a feature that, he lamented, the "painter hath not well expressed." Gerard wanted to accurately depict a plant, but was quick to critique the end result. It may seem obvious that an author who wishes to include illustrations would want to work with an artist themselves, but the real issue is that there are layers of unnecessary abstraction in this text. Abstraction from real plant to black-and-white illustration in Brunfels and Fuchs was just part of the process. In Gerard's case, the printer could have funded an original composition in accordance with the author's specifications, but instead chose to simply reuse Dodoens' illustration, itself a copy from an illustration that well pre-dated the turn to realism, forcing Gerard to point out its flaws. The temporal proximity of this text to those of Colonna, Alpini, and Donati demonstrates the transitional nature of the technique of printed botanical illustrations. Whereas continental European imprints were developing techniques for realistic images, especially in Italy and

Figure 5.19 *Coronopus ex codice Caesarea* (from a Dioscorides codex). John Gerard, 1636. Courtesy: Oak Spring Garden Foundation Library.

Germany, there were still holdouts that continued to employ the overly simplistic woodcuts that had defined early printed illustrations prior to Brunfels, in some cases simply reusing other woodblocks rather than commissioning new ones.

By the early eighteenth century, however, authors and publishers alike had realized that the audiences of these texts would not be satisfied with emblematic, simplistic illustrations. It was clear that while the authors themselves, and perhaps even their illustrators, may have had access to specimens with which to compare their textual descriptions, their readers likely did not. Accurate, detailed illustrations became paramount for those unable to travel overseas to obtain their own specimens or see the plants in the wild. This change can be seen in the depictions of ferns and fern-like plants in Charles Plumier's (1646–1704) *Traité des Fougères de l'Amérique* of 1705. Plumier frequently zoomed in on the unique physiology of the ferns, showing their vasculature, their lack of seeds and flowers, and the varying forms of their megaphylls (leaves). In this way, Plumier was able to depict the physical characteristics of an individual fern that distinguished it from other ferns and mimicked the increasing granularity of the study of plants and plant parts, thanks to the development of the microscope. In the case of a fern, which reproduces via spores, it is important to show what their megaphylls look like, rather than depict the entire plant. For example, the illustration of *Filix Ramosa* (plate XX2) shows a long, straight bipinnate frond stem with alternating stems at 45-degree upward angles (see Figure 5.20). Each of these stems is crowded with its own alternating pinnate leaf structures of acuminate shape, themselves consisting of alternating sporophylls, demonstrating mathematical precision and symmetry. While the sporangia in which the spores are housed cannot be seen in Plumier's illustration, the precision he was able to provide is nevertheless remarkable and serves as an excellent example of the turn toward scientific illustration. The audience for Plumier was very clearly not the herbalist

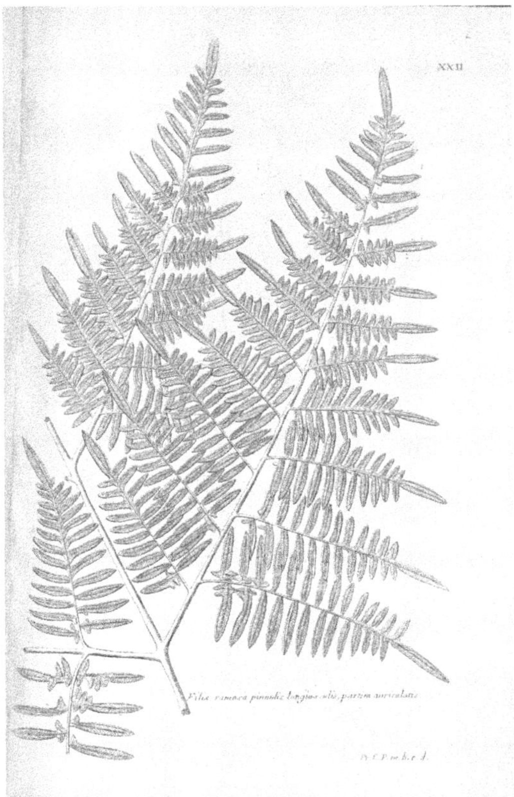

Figure 5.20 *Filix Ramosa*. Charles Plumier, 1705. Courtesy: Oak Spring Garden Foundation Library.

or doctor, but the botanist, amateur or otherwise, who was in need of a reference book on ferns in the Americas.[86]

Plumier's influence can also be seen in the classification and depiction of magnolias. In 1703, Plumier published his *Genera*, in which he first described a genus that he called *Magnolia* after the French botanist Pierre Magnol (1638–1715).[87] Linnaeus later cited Plumier in his *Species Plantarum* (1737), despite never having seen a specimen and having to rely on an illustration. The species that Linnaeus describes is *Magnolia virginiana*, although he may have confused it for the species described in Mark Catesby's (1683–1749) two-volume *The Natural History of Carolina, Florida and the Bahama Island*. Catesby illustrated most of the text, having taught himself to etch the plates. Catesby's illustrations display a similar level of precision as Plumier's, though with a bolder, brasher aesthetic. The illustrations are also in color, but it is possible to compare some of them with their black-and-white counterparts. In volume two of Catesby, published in 1743, the illustration of *Magnolia altissima* by Georg Ehret (1708–70),[88] a species now known as *Magnolia grandiflora*, is much larger than life and depicts a magnolia in full bloom, together with its fruit, a seed cone typical of species in the *Magnoliaceae* family (i.e., with seeds suspended from the cone). The bud is not yet florescing and some of the leaves are curled up on the edges to reveal the brown underside common to magnolias. All of this is on a black background, making the flower's features pop. In the copy owned by the Oak Spring Garden Foundation, a black-and-white printed broadsheet entitled *A Catalogue of American Trees and Shrubs* (1737) has been tipped in on page 61, opposite the color image. This image depicts the same species of magnolia in much the same manner, with very slight differences. It was engraved and etched on copper by Catesby for Christopher Gray (1694–1764) and measures 30 × 50cm (see Figure 5.21). The image is based on Ehret's illustration, while the English and French text is not specifically about magnolias, despite the dominance of the illustration. There are marginal notations that indicate preexisting texts in which the trees, alphabetically listed, are more fully described. The black-and-white illustration initially looks like a mirror of the color image, but there are several minor differences, in particular the absence of the bud. Despite this, Catesby's illustration manages to convey roughly the same information as Ehret's, especially the hermaphroditic features of the flower, including the stamens and carpels, the unique fruit, and the stiffness of the leaves.

Continuing the flora of the Americas, in 1758 Niccolò Gavelli (1701–77) published his *Storia Distinta e Curiosa del Tabacco*. The text has only one

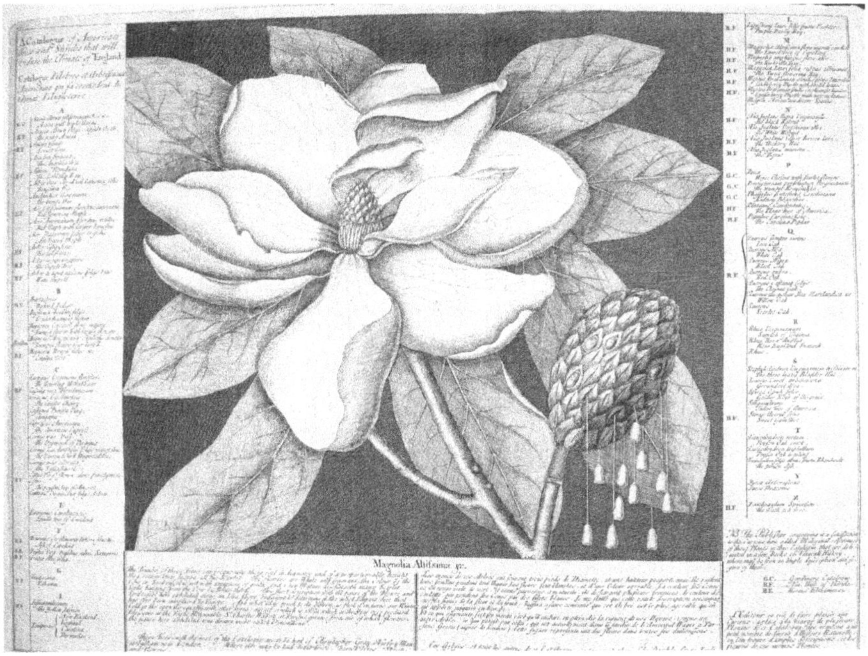

Figure 5.21 *Magnolia altissima*. Mark Catesby, 1737. Courtesy: Oak Spring Garden Foundation Library.

illustration of *Nicotiana Tabacco* on fol. 2r, a very simple woodblock that is nevertheless immediately recognizable as a tobacco plant due to the leaf and flower shapes (see Figure 5.22). Gavelli mentioned that *Nicotiana* was brought to France *c.* 1560 and included a couple of the local names for it, such as *Yoli*.[89] Just a couple of decades later in 1781, Giovanni Domenico Civinini's *Della Storia e Natura del Caffè*, also has only one illustration, inserted between folios 6 and 7 and entitled *Ramo della Pianta del Caffè* (Branch from a coffee plant) (see Figure 5.23). This illustration is quite detailed, showing on one branch the leaves with their pinnate venation in minute detail. It also depicts how the flowers emerge from the axil between the leaf and the stem, the location of the fruit, and even the palmately compound gathering of leaves at the apex of the branch. The artist also included three extra illustrations on the right side, one of the flowers in close up (*il suo fiore*); a second of the fruit (*frutto del caffè*); and a third of the seed (*il suo seme*), evidence of the changes occurring in scientific, botanical illustrations.

Only a few years prior, Étienne-François Geoffroy's (1672–1731) *Descriptions, Vertus, et Usages* (1767), a physician's herbal in five volumes, demonstrates how

Figure 5.22 *Nicotiana Tabacco*. Niccolò Gavelli, 1758. Courtesy: Oak Spring Garden Foundation Library.

the development of botanical texts and illustrations was circuitous, with frequent regressions back to the herbal tradition. Despite these relapses in the illustrations, the texts themselves would betray the influences of the developing field of botany. The first four volumes of Geoffroy depict plants, while the last focuses on animals. This is a herbal and not a botanical text whose author is concerned with being the first to describe in Latin or any Western European language newly "discovered" plants. Many of the herbs can be found throughout the herbal tradition, such as *Consolida maior*[90] and *Sanicula*,[91] which Brunfels had called *Solidago*. However, the entry on *Papaver hortense* (Opium),[92] which describes the medicinal benefits of the drug, is a thoroughly modern illustration, depicting the entire plant with skilled precision (see Figure 5.24). The lines of the leaf and petal venation, shadows, and texture of root, stem, leaf, and bud, are finer and

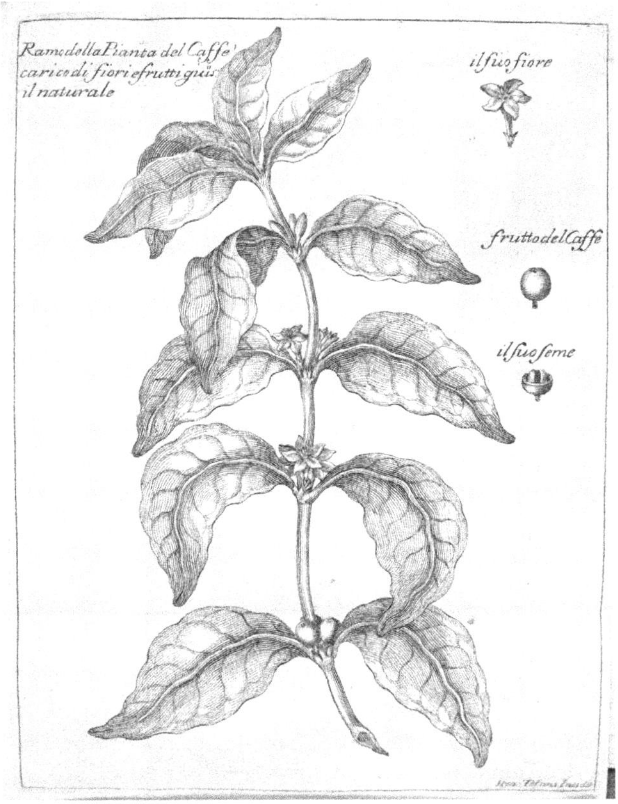

Figure 5.23 *Ramo della Pianta del Caffè* (Branch from a coffee plant). Giovanni Domenico Civinini, 1781. Courtesy: Oak Spring Garden Foundation Library.

more complex than in earlier woodcuts. Overall, the plant comes across as prickly and textured in ways that other illustrations cannot approximate. On the left is a close up of the fruit and to the right a close up of the flower in full bloom (florescence), a complement to the full illustration, which shows flowers in less mature stages.

Conclusion

The ideals of discoverability, knowability, and universal intelligibility are embedded in botanical Latin and nomenclature, but they are also features of botanical illustration, which has its own distinct history, entwined with that of

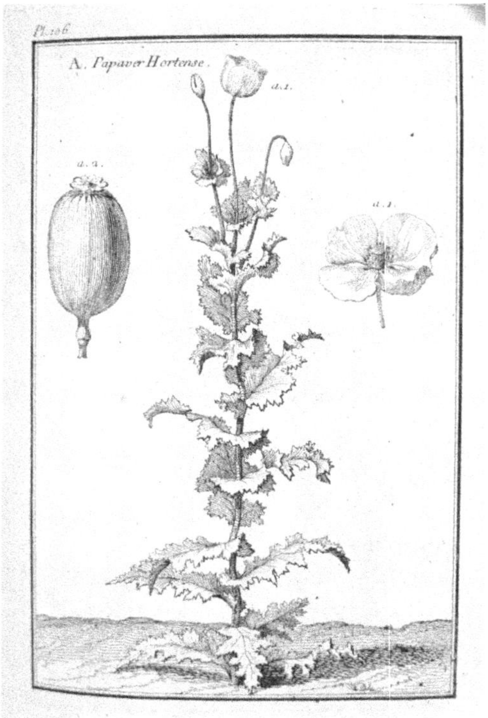

Figure 5.24 *Papaver hortense* (Opium). Étienne-François Geoffroy, 1767. Courtesy: Oak Spring Garden Foundation Library.

botanical Latin. Botanical illustration is also strongly aligned with the herbal and medicinal traditions, in which the issues of knowability, recognition, and intelligibility have practical, real-life implications: a mistake in identification of a plant used for a herbal remedy can have deadly consequences. The tradition of botanical illustrations runs so closely parallel to the development of botanical Latin that one cannot discuss the one without the other. Some botanists have even incorporated the drawings of plants into their formal training, precisely because a practitioner is more likely to depict plants without concern for artistic vision, which is better for conveying essential information about them. Some naturalists have even argued that an "illuminated" (colored) illustration of a plant can be better for understanding it than a textual description. Where a botanist cannot or is unwilling to produce their own illustrations, the second-best scenario is to work closely with a trusted artist, who understands what information the botanist needs to communicate to the scientific community. William Blunt (2021) considers botanical illustrations produced in such circumstances to be the most useful,[93] while Wittman (2013) points out that many scientific illustrators such as those employed by natural history museums, are not trained scientists but artists. These illustrators both supplement the scientist's observation and "defamiliarize" it so that the product of their own study of the organism has the ability to supplant the nomenclature and textual descriptions produced by the scientist.[94]

The value and efficacy of botanical illustrations is not a new issue. In the first century CE, Pliny the Elder associated them with the Greek tradition and discussed their flaws. He advised against using illustrations, arguing instead for descriptive names, textual descriptions, and direct access to and empirical observation of plants. By these means, he thought that plants could be known and discovered. The advent of printed illustrations using woodcut prints introduced an extra level of potential inaccuracy in terms of making a plant discoverable or knowable, beyond the mistakes of the manuscript tradition. In Pliny's time, and before the printing press, the image was original, which meant that its concord with the textual description was the responsibility of the artist. In Brunfels' time, the very processes of printing meant that even if the original artist rendered the plant perfectly, the final product might not accurately represent it. Despite the developments in printing techniques, and in particular the ability to print color illustrations, this issue of abstraction remains and is evident even in the most technical, detailed etching. Linnaeus himself was ambivalent about illustrations, at one point claiming they were useful only for fools and children. In his *Genera Plantarum*, he stated:

> I do not recommend images for the purpose of designating genera, but I wholly reject them and I must confess that they are better suited for children and those who have more head than they have brain.[95]

Yet in 1737 Linnaeus had published the *Hortus Cliffortianus*, which was illustrated by Georg Ehret and named after George Clifford, a Dutch banker with an English-style garden and herbarium in Hartecamp, who had commissioned and financed it. Despite this earlier collaboration, Linnaeus did not consider illustrations a necessity for his work on classification. Nevertheless, illustrations continue to be used by botanists, and newly published genera and species are almost always accompanied by illustrations of the various identifying parts. The ideals of universal intelligibility, discoverability, and knowability, are not only embedded in botanical Latin and nomenclature, but also in the choices of images accompanying the published text. Clarity and depth can be provided to the ongoing debates in the field of botany by focusing on the importance of illustrations for all three of these ideals. Illustrations do not carry the same baggage as language and text, which are more obviously aligned with and the result of colonization. Despite this, as the field of botany developed, so did the styles of illustrations and just as the Code dictates that "traditional knowledge" and indigenous names be relegated to protologues and not the main, scientific part of the publication, so too can we see botanical illustrations becoming more selective and exclusive in the information they convey. Medicinal uses, agricultural and natural historical contexts, environmental factors, textile applications, are all considered unnecessary for identification purposes and they eventually disappear entirely. But identification for a taxonomist is not the same as identification for a forager or herbalist, and the botanical illustrations that supplement the texts of the former are often of little to no use to the latter. In and of itself, this fact is not particularly problematic. The problem lies in claims to knowledge. Scientific illustrations convey highly specific information about plants, which is not the sum total of information about the plant. Those who have access to this other information are not unknowledgeable about the plant simply because the IBC does not consider it valid for publication.

Conclusion

What does it mean to know a plant? How does a plant's name, its textual description, or its visual representation work to render it knowable? Can any of these transcend culture and language to achieve universal intelligibility? These are the questions that have directed and informed the discussion at every point in the preceding chapters, which delineated the basics of botanical Latin, discussed its origins in antiquity, its modern iteration, and the development of certain principles in the field of botany itself and in its relationship to Latin as its language of expression. The ideals of discoverability, knowability, and intelligibility are identifiable in the works of Pliny the Elder, Otto Brunfels, and the sixteenth-century botanists whose work informed the taxonomic projects of Carl Linnaeus. The progress of the field from the eighteenth century until the present day is also characterized by an attempt to resolve these issues by various methods, first and foremost of which was the formal adoption of Latin for vocabulary, nomenclature, and diagnoses, and the informal adoption of certain styles of composition in Latin, specifically the heavy reliance on the ablative case.

These attempts were almost immediately succeeded by intense debates about their efficiency and ethics. The control wielded by the International Botanical Congress (IBC) has had twin effects, of ensuring that the publication of new organisms meets certain standards, and of forcing a particular format in a language that (inadvertently) excludes many would-be practitioners. Much as the printing history of Pliny's *Historia Naturalis* involved the establishment of an authoritative version that itself influenced Otto Brunfels and the discipline of botany, the International Code of Nomenclature for algae, fungi, and plants (ICN) has become the authority on plant naming and description. It does not, however, follow from this that the Code is the only way to name and know a plant or that any naming and descriptive practices that exist outside the Code do not have validity. To be fair, the aim of the Code is not discoverability, but to make as clear as possible the taxonomic rank of an organism, which is why the IBC is able to continuously sidestep the issue of including in a publication the non-diagnostic features of an organism. The medicinal benefits of a plant do not contribute to the taxonomy of the plant and they do not classify it. Neither do knowledge and recognition of indigenous names, or the colonial nature of a plant's "discovery," naming, and description. However, the field of botany is currently embedded in a discourse about these very issues precisely because science itself is on the cusp of a transformation. In particular, the past century has seen considerable discourse about and critique of scientific communication, scientific knowledge, and methodologies.

Scientific Knowledge and Methods of Knowing

In a 2016 follow-up to his two-volume *A Social History of Knowledge*, Peter Burke discussed several studies of the history of knowledge.[1] As an academic discipline, the historical approach to knowledge systems is relatively new, primarily developing in the 1980s and 1990s with, for example, B. S. Cohn's *Colonialism and its Forms of Knowledge* (1996) and R. D. Brown's *Knowledge is Power* (1989). This culminated in the twenty-first century with *Lieux de Savoir*, edited by C. Jacob (2007–11), and Burke's own works, which are focused primarily on Western Europe from the Early Modern era until the present "information age." More generally, scholarly interest in categorizing knowledge systems is hardly new. Francis Bacon had written about the topic as early as 1605 in *The Advancement of Learning*, where he argued for knowledge reform based on a systematic history of the various academic disciplines.[2] In the nineteenth century, the idea of constant forward progress in the development of knowledge systems and acquisition was ubiquitous, even informing Darwin's *Origin of the Species* (1858). At the same time, there arose an interest in the social contexts of knowledge, as seen in both Karl Marx's critique of knowledge as a product of social class, and August Comte's doctrine of positivism, which states that all knowledge is either axiomatic or *a posteriori*, derived from deductions based on logic and reason. Positivism itself Comte viewed as the final, cumulative stage of the history of human knowledge, the first two being theological (fictitious) and metaphysical (abstract).[3] The study of knowledge as an amalgam of, or tension between, science and society culminated in the twentieth century in a *Wissensoziologie* (sociology of knowledge), a loosely organized discipline focused on what people know, how, when, and in what social contexts. It is often credited to Karl Mannheim (1893–1947), whose *Ideology and Utopia* broadens the Marxist idea that a social group's ideology is a result of their social class and condition to the proposition that an individual's beliefs are also the product of their social-cultural context.[4] In the *Wissensoziologie* tradition, general histories of knowledge, the social sciences, and humanities have all been modeled on the history of the natural sciences.[5] Burke sees the current "digital revolution" as one in a series of historical shifts in knowledge production, but views the history of knowledge as the product of two even more specific individual histories, that of the book (its economic history as well as its cultural impact) and that of science.[6]

Throughout this book, the scientific field of botany has been shown to have origins in the history, transmission, and reception of a particular text, Pliny the Elder's *Historia Naturalis*. However, rather than specify the exact relationship between book history, general knowledge, and the history of science, it may be more helpful to envision the various points of influence as linked data, where one reference sphere, such as the first printed edition of Pliny, has links to any number of other points, such as the late twentieth-century iteration of botanical Latin. Depending on the direction one chooses to focus attention, that reference sphere may become another reception sphere. There is always, in this model, potential for any point to be related to another and to serve as both reference and response, to transform and be transformed. Either way, whether we evaluate the history of botany on Burke's model or via transformation methodology, the history of knowledge and knowing informs the legacy of botanical Latin. Critiques of knowledge production provide a framework for understanding the history and current status of botanical Latin. A distinction must be made, however, between modern conceptions of science, which originate in the nineteenth century, and earlier systems of knowledge production. Botany is modern science and is therefore subject to a narrower set of methodological constraints and a more specific set of axioms than the sixteenth-century botanist-herbalists. The term *scientia* simply means "knowledge" in Latin and it is for this reason that we tend to add descriptive modifiers to the term in order to distinguish between the hard sciences, social sciences, humanities, medicine, and technology. Despite the differences between these disciplines and their objects of study and methods, the same or similar critiques have been made of them.

In 1962, Thomas Kuhn discredited the illusion of steady, linear, progress, positing instead the paradigmatic nature of scientific progress. Kuhn rejected the idea of scientific "development-by-accumulation"[7] and described a circular model onto which one can map the stages of a particular paradigm: from pre-scientific to "normal science," to crisis and revolution, from which a paradigm shift results. Vickery (2000) argues that in the twentieth century, a "pervasive feature" of science has been "uneven development," although he cites the world of business as the primary example of this feature.[8] Meanwhile in the literary sphere, Michel Foucault's archaeological method of historical analysis assumed a fundamental discontinuity between epistemes[9] and a structural disunity between text and reader that precludes any prior or mutual understanding.[10] Historical "knowledge" is essentially just an amalgam of points on a continuum of constantly changing discourses, which he calls "discursive formations," in which it is possible to identify themes, but not "meaning" in the sense of objective, ontological truth. The semantics assumed by Foucault amount to the idea that an expression or proposition must be evaluated by reference to the conditions in which it arose and in the context of its domain of discourse. The progression from one set of rules in an episteme to another Foucault calls "discontinuity," since in a very short period of time, a culture might switch from one belief system, ideology, or way of thinking to another.[11] Foucault was influenced by the discourses of Mikhail Bakhtin's dialogic principle, which assumes a polyphonic discourse in texts, characterized by fundamental disunity and a lack of harmony. Within the tradition of early twentieth-century critical social theory, the opposite is usually assumed, namely, that the relationship between self and other is defined by a fundamental unity and shared understanding, Yet even Jürgen Habermas pointed out that "collectivities" remain integrated only when the individuals that populate them have sufficiently communal and reciprocal beliefs that resolve into uncontested "background convictions."[12] In social-literary contexts, there was a turn away from hermeneutics, ontology, and axiomatic principles, toward phenomenological critique. As a trained scientist, Kuhn continued to operate under the assumption that there was such a thing as scientific knowledge, but his critique of the methods of science are compatible with those of the literary theorists of his time.

One of the most serious methodological critiques of the hard sciences involves the replication crisis and scientific studies with unreproducible results.[13] The axiom of replication assumes the feasibility of and need in experimentation for a vacuum-like environment devoid of as many outside variables as possible. The critique of the replication standard therefore often focuses on the implications of the inevitable influence of subjective processes and cultural contexts on scientific investigation and "knowledge." Simply put, science does not actually exist in a vacuum and ethical concerns not only do, but should, influence the ways in which science is done. Despite this, some, such as Daston (2008), have argued for a return to ontology in scientific observation. By ontology, Daston does not strictly mean metaphysics, but rather, the stabilization of the objects of scientific observation so that the very focus of scientific study is actually appropriate for the methods of science. Daston claims that this ontology lies between metaphysics and epistemology, but it is also aligned with reverse phenomenology and transformation methodology, both of which place equal focus on the biconditionality of observer and observed. Daston is concerned with the scientific domain of discourse, arguing that ontology is focused on the ways in which scientists essentially construct a universe composed of objects that can be observed using various established methods of scientific investigation, but which "rarely correspond to the objects of everyday perception."[14] It is both a highly focused definition of what constitutes an appropriate object of scientific observation and an acknowledgement that some phenomena cannot be known scientifically. For Daston, this recognition is meant to resolve the epistemological issues plaguing science, but it also opens discussion about the limitations of science. For organisms that cannot be easily observed with the naked eye, there is a disconnect between scientific classification and observation. Unlike in botany, some physical, observable phenomena, such as cloud formations, are far too varied and mutable to easily correspond to the taxonomy and nomenclature of scientific researchers.

Broadly, phenomenology is concerned both with phenomena, the objects of our experience, and the kind of investigation into phenomena that originates from an inner stance, on the meaning of the objects within one's own lifeworld. There are other ways of defining phenomenology, but a fundamental criterion is that the phenomena, the experiences in question, are intentional. The direction of focus is one-way, from the subject to the object so that the relationship between the two itself constitutes a phenomenon. Phenomena and experiences range from emotions to thought, from logic to imagination, from sound to silence, since even absolute silence is a perceptible situation that one experiences. But phenomenology also has the potential to be reciprocal, such that the element of intentionality can switch direction at any moment, allowing ideas to be communicated and directed at an individual or a group of people, creating an interpersonal, interconnected super-phenomenon. Possible, and perhaps even inevitable, effects of these intertwined phenomena include miscommunication, misinterpretation, and a semantics that is not static but dynamic, difficult if not impossible to pin down and capable of changing at any moment. Bakhtin picked up on this dynamic semantics in his definition of the pre-novel genre *menippea*, crucial to which is the role and perspective of the author, where the narrator "discovers that he means more than one thing" and "truth comes as a possibility that arises from the conflict of ideas."[15] However, Bakhtin also concluded that the principle that defines *menippea* is "carnival and a carnival sense of the world."[16] The idea of the "carnival" is in many respects related to societal otherness, to a celebration of the underbelly of human civilization and a reveling in those aspects of human existence that make us most uncomfortable. Bakhtin might have characterized this reciprocal phenomenological phenomenon as carnivalistic. For scientific knowledge and semantics, the implication of this critique of texts and directions of observation is that what constitutes knowledge and truth turns out to be considerably less stable than the sciences tend to admit. The assignment of a semantic value to a statement or proposition can be understood as a process that is fundamentally unstable and capable of sliding at any moment in either direction, toward truth, or toward falsity, or somewhere in between.

Visual Communication and Knowledge

Knowledge about plants is communicated textually, orally within local traditions, or visually. Botanical illustrations have developed in tandem with the science of botany and have evolved to convey very specific and precise information about plants. It is tempting to treat images as more universal than language and to a certain extent, this is true. An individual need not share a language with the artist in order to recognize basic information about a plant, such as whether it is a tree or a shrub, a flower or a vine. It is also true that a plant illustration made or commissioned by a plant scientist is "speaking" to a very narrow and particular audience and is therefore communicating with them alone. It is conveying diagnostic information about the plant to the exclusion of other information. If a bystander without the same shared scientific training happens to encounter the illustration, they are almost certainly going to read it differently and take different information from it. This is as unavoidable as a text being read in wildly different ways by different people in different time periods. At the same time, the very act of creating a visual mode of communicating information affects the way in which the scientist parses and evaluates the data they are collecting, both in the moment of observation and in the data analysis stage. What we know or think we know about the physical world is informed by how we interact with it. In recent decades, the connection between visual communication and knowledge, long known to botanists, has been discussed as a crucial component in scientific methodology. Drawing on mathematical notions of fuzziness, Jordi Cat discusses what he calls "pictorial empiricism," a methodology that not only incorporates images into scientific observation, but also embraces the idea that in the sciences, "fuzziness is widely unavoidable and unreasonably effective."[17] Cat's argument is that visual imprecision and incompleteness mirror our actual experience of the external world, which we inevitably see in fuzzy ways. This idea is extensible and can be applied to botanical Latin and nomenclature. It also applies to the scientific problem postulated and presumably solved by Linnaeus' axiom that there must be a 1:1 ratio between plant and name. Just as an illustration of a plant is a reflection of the implicit biases and motives of the observer, so too is human language dynamic and diverse. There will never be one name for a plant that is universally intelligible, any more than one illustration will correspond perfectly to the expectations of every reader.

The goal of many recent critiques of the sciences seems to be a kind of methodological revision, focused on the ways in which observer and observed interact. Similarly, the point of this book is a critique of the very axioms of science and their embodiment in ideals like universal intelligibility, using the narratives of botany and botanical Latin as mediums. The scientific method is based on a particular system of logic, known informally as classical logic because of its origins in Aristotle and the Renaissance, and formally as first-order logic. The system is robust and consistent, consisting of a semantics with which to determine the truth and falsity of propositions,[18] and a set of axioms and rules of inference and deduction with which to determine the validity of an argument. Despite the strength of this system, it is limited in scope, unable to handle any philosophical issues that stand outside of its domain of discourse. In particular, first-order logic takes as an axiom the law of bivalence, according to which every proposition has exactly and only one of two possible truth-values (true or false), represented numerically as 1 and 0 and syntactically as the law of the excluded middle (P or not-P). A second, related axiom is the law of non-contradiction, according to which a proposition cannot be both truth and false, represented syntactically as [not (P and not-P)]. There are, however, other non-classical systems of logic, including many-valued logics, where the number of possible truth values range from three to the set of all real numbers; or discussive logics, where the domain of discourse allows a proposition P to be asserted by one participant at the same time as its contradiction ¬P is asserted by another, rendering the law of non-contradiction no longer relevant or axiomatic.

For instance, in 1963, Jan Łukasiewicz proposed what he called "many-valued logic."[19] Regarding the law of non-contradiction, since sentential logic assigns the value "true" to any sentence or proposition that can be proven via the rules of inference, if two contradictory statements were thus proven, each would have to be taken as true.[20] But this would render the entire system inconsistent, which would further mean[21] that any and every statement in that system of logic that has meaning becomes a thesis. As a thesis, it has truth value and if every meaningful statement has truth value, then the distinction between truth and falsity disappears.[22] Łukasiewicz's goal was to prove that the system is in fact consistent and that its axioms are independent of each other.[23] He accomplished this by using a third truth value, 2, over and above the 0 used for falsity and 1 used for truth. The numeric value of 2 is beside the point; what matters is that one can posit a third value. The symbols 0 and 1 are not actually referring to true and false sentences, but if they can be interpreted logically as functions, then the third value must be as well. A sentence can therefore have a truth value other than false or true, or, if the truth value is unknown, it might have no truth value, or a third one besides true or false. Łukasiewicz also posited that between 0 (false) and 1 (true) there could be an infinite number of values, much like probability calculi.[24] He therefore allowed for 3-valued logics (0, 0.5, 1), 4-valued, etc., all the way up to the set of rational numbers x between 0 and 1, where $0 \leq x \leq 1$. Łukasiewicz posited a range of semantic values that is defined by rational numbers; however, in 1965, Zadeh tweaked this to include the set of all *real* numbers between 0 and 1, inclusive. This renders the set finitely infinite, since real numbers can be irrational in addition to rational. This also has the effect of rendering propositions mappable that would otherwise be too inexact. With an infinite continuum, a proposition need not be completely false or completely true, but can be located anywhere in between. Crucially, this model relies on a notion of vagueness, as opposed to the unknowability of probability theory. If a proposition's truth value is an irrational number like $\sqrt{2}$, the presumption is that it has its origin in an inexact or incomplete process of reasoning.

Łukasiewicz and Zadeh were working in the context of mathematical logic, but their ideas can be extrapolated out into the scientific method, which assumes only two truth values and has no room for vagueness or inexact propositions. Bruno Latour (1990) argued against the grandiose dichotomy of the "scientific" versus the "pre-scientific," a division that assumes some undefined fundamental change in the human mind, which is not supported by the history of human thought. For Latour, the conventional assumption of this dichotomy serves as an explanation for, and could even be the cause of, the vast differences in scientific effects. He argues instead for a fully empirical, but not relativist, account of scientific knowledge that focuses on "writing and imaging craftmanship."[25] The twin methods of writing and drawing, he says, collapse the lofty theoretical systems and false dichotomies they assume down into a methodology in which individuals engage in discourse and argumentation with more than just words. They engage each other with "paper, signs, prints and diagrams."[26] Hoffmann and

Wittmann (2013) argue that scientific illustrations, in conjunction with text, convert scientific concepts into something unambiguous and intelligible. In the process, illustrations "shape and develop research questions" and are therefore an essential part of knowing.[27] They concretize the data, render it familiar and approachable, and stabilize it as a holotype that guides the reader. Gross and Harmon (2014) note that the sciences rely on visual communication far more than literary studies, because scientific truth "almost always requires making something visible and analyzable."[28] From this, it follows that "scientific meaning" is the result of a collaboration between texts, images, and tables.[29] This is the case for all scientific disciplines, the visuals of which changed over time until they ceased to depict the natural world in the way an artist would. Scientific illustrations transformed into something that could be measured and subjected to the axioms of mathematics and science.[30] Scientific illustrations are quite literally scientific arguments and are therefore quantifiable in ways that an artistic illustration is not. Gross and Harmon believe that the "heart of scientific communication is scientific argument." If so, scientific illustrations, botanical ones included, are visual forms of a scientific argument.

Intelligibility

Alix Cooper (2009) has observed that compilations of indigenous plants in Europe were happening simultaneously with the collecting of specimens from abroad. The recognition of the value of local plants and the acknowledgment of them as indigenous was informed by European botanical expeditions in the Americas, Africa, and Southeast Asia. This realization corresponds to the ancient debates about simple herbs versus complicated, foreign medicaments, but it also parallels a modern issue. Plant scientists have now described the near-entirety of the plant kingdom, even DNA sequencing organisms, and this knowledge has been compiled in countless databases and inventories. With much of this information readily available online, new questions arise with regard to local and indigenous plants. What do we do with the vast quantity of information we have? It is possible to answer to this question strictly scientifically, but it also has a moral resolution. Latinized names forced on indigenous plants must be deprivileged and the implicit assumption of universal intelligibility must be reconsidered. This does not mean throwing out all of this collected knowledge, but simply rejecting the assumption that to know a plant is to know it in the ways these repositories assume. European colonizers did not just ignore indigenous plant names. Were it that simple, one could argue that the legacy of botany is simply one of disdain or indifference at best. As Wendy Makoons Geniusz (2009)[31] points out, the "colonization of botanical knowledge" was also characterized by the appropriation of indigenous plant knowledge. Europeans did in fact learn from the indigenous people of the Americas, and they did so in a reflexive way. "For themselves" is a short, but powerful phrase. They learned medicinal benefits of plants endemic to other lands, plants that were useful for survival in those lands, and they not only took it for themselves, they also, insidiously, completed the colonization process by convincing indigenous people that their own knowledge was primitive and inferior. Believing this to be true, many tribes willingly gave up their knowledge, first for pay, and later for the myth of assimilation into the culture of the conquerors. Geniusz herself was raised in Ojibwe language and culture and her methodology involves presenting reclaimed *anishinaabe-gikendaasowin* (anishinaabe botanical knowledge), originating from the various tribes known broadly as the *Anishinaabeg*, such as the Ojibwe, Ottawa, and Potawatomi. Part of this process involves providing the *Anishinaabemowin* or *Ojibwemowin* (Ojibwe) names of plants, which are listed first, with their Latin nomenclature in parentheses. This is a very purposeful and astute decision, since it overtly deprivileges the scientific names in favor of their Ojibwe names, but it also calls attention to the colonization of knowledge of the plants. Illustrations of plants are accompanied by Ojibwe songs and traditional stories and knowledge of the plants. Anticipating the potential criticism that her own work is inadvertently colonized, since it is a published text, Geniusz provides a two-part definition of colonized texts. For a text to be colonized, it must either serve the "interests of the colonizers and the processes of systemic racism and oppression," or its material and the information contained in it must align with colonizing ways of thinking and knowing, which are at odds with those of the *anishinaabe-izhitwaawin*.[32] Similarly, Frantz Fanon described the decolonization process in 1963 by focusing on the "colonized subject," who is forced to exist in a defensive position, bewildered by the colonists' cultural impositions. The colonized individual is unsure of their place in the colonists' world, and while they know they are viewed as subaltern, they are not persuaded of their supposed inferiority.[33] Despite the colonizers' attempts to suppress the spirit, culture, beliefs, and language of the colonized subject, they are never entirely successful. The methods and systems of knowledge that belong to the oppressed do not entirely die off, they simply become dormant. It is in reference to this that Smith (2012) makes careful use of the terms *re*present and *re*search to note of the imperialist agenda that "knowledge was also there to be discovered, extracted, appropriated and distributed."[34] This indigenous knowledge was *re*presented to the West by Westerners and indigenous people themselves had to *re*search their own cultural, historical knowledge through the systems of classification, naming, preservation, and investigation with which Westerners chose to understand this stolen knowledge.

By replacing vernacular names with Latin ones, which Linnaeus formalized in the eighteenth century, naming and referring to plants became a colonial act. Formal, scientific Latin plant names are only "known" to plant biologists and to amateur botanists or professional gardeners who have taken the time and effort to memorize them. The common names by which people have long referred to plants have been deprivileged with the result that "knowledge" of botanical nomenclature is fundamentally exclusive.[35] But it does not follow from the fact that if an individual does not know the formal scientific name of a plant, they do not know the plant itself. The changes currently underway in the field of botany are part of yet another transformative process, the result of which will involve a turn away from Latin nomenclature and toward an understanding of knowledge and knowability that acknowledges the limitations of the ideal of universal intelligibility. The arguments put forth against honorifics and for reversion to indigenous names do not assume this ideal. Rather, they assume a conception of knowability and the transmission of knowledge about plants that is moral, not scientific. Modern botanists, in pushing back against the rules for expression in their field, have therefore come full-circle, wrestling with similar philosophical issues as did Pliny the Elder and Otto Brunfels in 1530.

A plant name does not need to be universally intelligible in order to be discoverable or knowable. By removing the Latin requirement for publication in 2011, the IBC recognized this fact for scientific diagnoses. The same issues apply to botanical Latin nomenclature. If any degree of intelligibility is to be achieved, botanists will need to go further than proposals to reject particular names. They will need to consider what it would take to de-Latinize their entire system of nomenclature and whether or not doing so will resolve the current impasse.[36] This will also constitute a new transformation of botanical Latin, in which the limitations of universal intelligibility are recognized, namely, that knowledge of a plant is not dependent on any particular system of nomenclature. If we return to the plant *Gauchblům* and reconsider it in light of the ideal of universal intelligibility and the botanical community's continued adherence to Latin binomial epithets, we can see the philosophical issue at hand. Brunfels himself knew the plant *Gauchblům*. He knew what it looked like, what its medicinal properties were, and how it grew. The *vetulae* (old women) and *herbarii* (apothecaries) of Germany also knew the plant. None of this is negated by the fact that it had no known Greek or Latin name. Fuchs may have identified it, but even that name fell into disuse. *Nasturtium agreste* is no longer a recognized formal botanical name. Linnaeus recategorized it in 1753 as *Cardamine pratensis*, although he did not describe it. Rather, he simply noted its earlier names, including those provided by Bauhin.[37] He did not mention either Fuchs or Brunfels, despite the fact that it was the latter's frustration with identifying it and the former's attempt to do so that led to its classification and description in later works.

If the field of botany eventually does away with Latin nomenclature, it will be left with an extensive catalog of names that need review, some of which may revert back to indigenous and vernacular names. If so, the influence of Brunfels' *herbae nudae* on the botanists who succeeded him will become even more apparent. Had Brunfels been able to identify them and provide them with Latin names, they might still have been unintelligible to many people. The formality of their Latin names could in fact have rendered knowledge of them inaccessible to the very people to whom Brunfels owed much of his own *cognitio* (knowledge) about plants. It is my contention that botany as a discipline instituted such strict rules for publication because those rules were meant to be exclusionary from the very beginning. Plants are tied to medicine, textiles, food, and other crucial aspects of life, and therefore to power. As botanists from Western Europe had to look further and further abroad in order to find plants that they could publish as "new" and therefore make themselves authorities and gain recognition, it became expedient to restrict the language of communication to Latin, which automatically excluded most people in the world.

Notes

Introduction

1 Carl Linnaeus, *Genera Plantarum* (Paris: M. A. David, 1743), iii. The use of *cuicunque* could certainly be taken as "to anyone trained in the field of botany" rather than "to any one person on the planet." However, I do not think that this is Linnaeus' point, nor was it Pliny the Elder's.

2 The original code was called either the *International Code of Botanical Nomenclature* (ICBN) or the *International Rules of Botanical Nomenclature* (IRBN). Earlier versions are specified by these abbreviations.

3 The Cambridge Code was not published until 1935 and is therefore often referred to as the 1935 Code.

4 John McNeill, "Latin, the Renaissance Lingua Franca, and English, the 20th Century Language of Science: Their Role in Biotaxonomy," *Taxon* 46 (1997): 753.

5 William Stearn, *Botanical Latin: History, Grammar, Syntax, Terminology and Vocabulary* (London: Nelson, 1966).

6 Stearn, *Botanical Latin*, 3.

7 Stearn's text is monolithic. The only real contender is Andrei Baranov, *Basic Latin for Plant Taxonomists* (Lehre: J. Cramer, 1971). More broadly, Epstein et al., *The Latin of Science* (Mundelein, IL: Bolchazy-Carducci Publishers, 2019) provides to readers scientific passages from antiquity to the Renaissance, along with commentary, vocabulary, and syntax notes and is intended as a reader for intermediate students of Latin. Interestingly, botany is not one of the scientific categories of the book, although medicine is.

8 Those in favor of retaining the Latin requirement included Filgueiras (1997) and Kabuye (1990). In favor of replacing it with English were McNeill (1997), Kostermans (1990), and Chaudhri (1991, 1992). This debate is discussed in greater depth in Chapter 4.

9 The most recent, published Code (International Association for Plant Taxonomy, Madrid Code, 2025) may be found here: https://www.iaptglobal.org/_functions/code/madrid.

10 Emma Short and Alex George, *A Primer of Botanical Latin with Vocabulary* (Cambridge: Cambridge University Press, 2013).

11 Ross Bayton, *The Gardener's Botanical: An Encyclopedia of Latin Plant Names* (Princeton, NJ: Princeton University Press, 2020).

12 Peter Jones, *Quid pro Quo: What the Romans Really Gave the English Language* (London: Atlantic Books, 2016), 311.
13 Lorraine Harrison, *Latin for Gardeners* (Chicago, IL: University of Chicago Press, 2012).
14 John Wright, *The Naming of the Shrew* (London: Bloomsbury Academic, 2014), xi.
15 Stearn, *Botanical Latin*, 11.
16 Binomial nomenclature is not unique to botany. It is the standard format in many scientific disciplines, as is a governing body that produces and enforces rules for nomenclature. There are some fields that do not have such oversight, however. For example, in entomology, while nomenclature follows the two-name format, there is very little oversight, resulting in such insect names as *Ba humbugi* (a land snail from Fiji); *Aha ha* (a species of Australian moth named after the exclamation A. Menke made when he opened a package containing a specimen); *Agra vation* (a carabid beetle); *Eubetia bigaulae* (a tortricid moth, pronounced "You betcha, by golly"); and so on. The only thing these names have in common with botanical names is the binomial format and the fact that each name is unique to exactly one species. There is no attempt to Latinize many of the names and in some cases, an English word has simply been split into two. Thanks to Benjamin Fortson for bringing to my attention these examples.
17 Stearn, *Botanical Latin*, 6–7. This is a summary of Stearn's definition.
18 For example, *Humulus lupulus*, the formal name for hops.
19 For instance, he discussed *Heracleon siderion* (25.15); *Helleborus candidum* and *nigrum* (25.21); *Hiera botane* (25.59); *Astaphis agria* (23.13), and many others. It is important to note, however, that Pliny did not use binomials exclusively or even with any consistency. Dioscorides also used binomials in *De Materia Medica*, discussing in Cap. VI two kinds of *Nardus*, one from India and the other from Syria: *Nardus indica* and *Nardus syriaca*. However, Dioscorides wrote in the same period as Pliny and there is so much overlap between the information in their texts that it is entirely possible that Dioscorides adopted binomials from Pliny.
20 Or a *differentia specifica*, in which, instead of a species name, the genus name would be followed by a diagnostic phrase, which was only required for more than one species. This is discussed in August Quirinus Rivinus, *Introductio Generalis in Rem Herbarium* (Lipsae: Christoph. Güntheri, 1690), 15.
21 *Historia Naturalis*, Book 25. This is discussed more fully in Chapter 1.
22 Carl Linnaeus, *Critica Botanica* (Leiden: Apud C. Wishoff, 1737), 2.
23 Linnaeus, *Genera Plantarum*, iii.
24 Linnaeus, *Genera Plantarum*, iii. I take the adjectives *similia* and *dissimilia* as pure substantives standing in for the missing term "things," not "plants," because I take Linnaeus' claim to be about more than just an understanding of how to think about the latter.

25 Ibid.

26 Stearn, *Botanical Latin*, 46.

27 Craig Kallendorf, *Printing Virgil: The Transformation of the Classics in the Renaissance* (Leiden: Brill, 2020), 1.

28 Kallendorf, *Printing Virgil*, 4. See also Lutz Bergemann et al., "Transformation: A Concept for the Study of Cultural Change," in *Beyond Reception: Renaissance Humanism and the Transformation of Classical Antiquity*, ed. and trans. Patrick Baker, Johannes Helmrath, and Craig Kallendorf (Berlin: De Gruyter, 2019), 9.

29 Kallendorf, *Printing Virgil*, 4.

30 Michel Foucault, *Les Mots et les Choses: Une Archéologie des Sciences Humaines* (Paris: Gallimard, 1966), 140.

31 Pamela Smith, *The Business of Alchemy: Science and Culture in the Holy Roman Empire* (Princeton, NJ: Princeton University Press, 1994).

32 Bergemann et al., "Transformation," 10.

33 Elizabeth Eisenstein, "A Classical Revival Reoriented: The Two Phases of the Renaissance," in *The Printing Press as an Agent of Change: Communications and Cultural Transformations in Early Modern Europe*, vol. 1 (Cambridge: Cambridge University Press, 1979), 168.

34 Kallendorf, *Printing Virgil*, 5.

35 Mikhail Bakhtin, *The Dialogic Imagination* (Austin, TX: University of Texas Press, 1981).

36 Michel Foucault, *The Order of Things: An Archaeology of the Human Sciences* (London: Tavistock, 1970), 326.

37 Bergemann et al., "Transformation," 11.

38 Craig Kallendorf, *Printing Virgil*, 5.

39 Craig Kallendorf, *Printing Virgil*, 8.

40 Mathura Umachandran and Marchella Ward, "Towards a Manifesto for Critical Ancient World Studies," in *Critical Ancient World Studies: The Case for Forgetting Classics*, ed. Umachandran and Ward (London: Routledge, 2024), 3.

1 Pliny the Elder's *Historia Naturalis*: The Origins of Botanical Latin and Nomenclature

1 The actual dates of composition are unverifiable, but he may have finished the first ten books by 77 CE, with the remainder unfinished at the time of his death in 79 CE, in which case the whole was published posthumously. This is plausible, since according to *Epistles* XVI and XVII of his nephew Pliny the Younger, he had sailed straight from Misenum to Stabiae when he heard of the eruption of Mt. Vesuvius, where he perished. It is very much in keeping with his profession to want to see the

eruption firsthand and incorporate vulcanology into his encyclopedia, but it is unknown if he was still working on it at that time.

2 A. G. Morton, *History of Botanical Science: An Account of the Development of Botany from Ancient Times to the Present Day* (London: Academic Press, 1981). Morton argued that the discipline began in 1483, when Theodore of Gaza, who had already assisted with the 1470 Roman edition of Pliny, translated Theophrastus' two botanical texts, *Historia Plantarum* and *De Causis Plantarum* into Latin. By doing so, Morton believes that Theodore reintroduced to Europe antiquity's highest botanical accomplishment at precisely the right time, when it was still "scientifically useful."

3 The famed *horti* of Rome, Pompeii, and Herculaneum were well established by Pliny's time, with the adoption in many elite households of the peristyle, which allowed for courtyard garden spaces. With the expansion of the empire and the accompanying land grabs, sprawling villas with gardens also became common. Especially in Rome, where the poor lived in *insulae*, access to such spaces was reserved for the elite and their households. Pierre Grimal, *Les jardins romains à la fin de la République et aux deux premiers siècles de l'Empire* (Paris: E. de Boccard, 1943), 290 provides a comprehensive overview of the style and history of Roman gardens.

4 For example, the *Medicina Plinii* of the fourth century is a compilation of the medical information in Pliny, whereas Solinus' *De Mirabilibus Mundi* extracts geographical information from the *Historia* and adds original material to it.

5 Sorcha Carey, *Pliny's Catalogue of Culture: Art and Empire in the Natural History* (Oxford: Oxford University Press, 2003), 18. Carey places the universal nature of the text against earlier writers, such as Varro's *Antiquitates* and *De Re Rustica*, and Celsus' *Artes*, arguing that the very structure of the *Historia Naturalis* reflects its encyclopedic function, beginning with the *mundus* and narrowing in focus down to the substance of the earth itself.

6 R. French, "Pliny and Renaissance Medicine," in *Science in the Early Roman Empire*, ed. R. French and F. Greenaway (London: Croom Helm, 1986), 252. The beginning of the end was Niccolò Leoniceno's 1492 tract, *De Plinii et Aliorum in Medicina Erroribus*, in which he listed the many scientific errors in the *Historia Naturalis*. For French, Leoniceno was the first to question Pliny's authority and not simply to lament the quality of the printed editions of his text. See also F. H. Garrison, "Herbals and Bestiaries," *Bulletin of the New York Academy of Medicine* 7 (1931): 892. Garrison claims that Aristotle and Theophrastus were the first to do "scientific" natural history and were unsurpassed until the Renaissance. Garrison does not deem either Dioscorides or Pliny true natural history authorities.

7 Eduard Norden, "Sein Werk gehört, stilistisch betrachtet, zu den schlechtesten, die wir haben," in *Die antike Kunstprosa vom VI. Jahrhundert v. Chr. bis in die Zeit der Renaissance* (Leipzig: Teubner, 1898), 1.314.

8 F. R. D. Goodyear, "Technical Writing," in *The Cambridge History of Classical Literature, vol. 2: Latin Literature*, ed. E. J. Kenney and W. V. Clausen (Cambridge: Cambridge University Press, 1982), 670.
9 Garrison, "Herbals and Bestiaries," 893. He says that Pliny is rife with "errors and superstition."
10 Michel Foucault, *Les Mots et les Choses: Une Archéologie des Sciences Humaines* (Paris: Gallimard, 1966), 130–3.
11 For example, Arist. *Resp.* 477a7, al: "Investigations into Living Things."
12 Andrew Wallace-Hadrill, "Pliny the Elder and Man's Unnatural History," *GAR* 37 (1990): 80–96.
13 Pliny did not specifically address the medical debates in Rome in the first century CE. However, Asclepiades was a Methodist, the only school of medicine that developed in Rome and was not of Greek origin, unlike Empiricism or humoral theory. The techniques of the Methodists were heavily criticized by Galen and others who worked in the Greek tradition.
14 A. G. Morton, "Pliny on Plants," in *Science in the Early Roman Empire: Pliny the Elder, His Sources and Influence*, ed. R. French and F. Greenaway (London: Croom Helm, 1986), 88–9.
15 J. N. Adams, "Introduction," in *Aspects of the Language of Latin Prose*, ed. T. Reinhardt, M. Lapidge, and J. N. Adams. Proceedings of the British Academy, 129 (Oxford: Oxford University Press, 2005), 4. Adams defends Pliny against the accusation that by sharing such constructions with poetry, he was trying to lend a poetic flavor to his writing. For example, the *dativus auctoris* is found in both Roman poetry and in Pliny. By not using a preposition, it not only subtracts from the word total, but also places more emphasis on "content" words than on "function" words. See also H. Pinkster, "The Language of Pliny the Elder," in *Aspects of the Language of Latin Prose*, ed. T. Reinhardt, M. Lapidge, and J. N. Adams. Proceedings of the British Academy, 129 (Oxford: Oxford University Press, 2005), 248.
16 Aude Doody, *Pliny's Encyclopedia: The Reception of the Natural History* (Cambridge: Cambridge University Press, 2010).
17 Ralph Jackson, *Doctors and Diseases in the Roman Empire* (London: British Museum Publications, 1988).
18 Vivian Nutton, *Ancient Medicine* (London: Routledge, 2012).
19 *HN* 25.10.
20 It is also tempting to take this proclamation as tongue-in-cheek, given that Pliny did not even provide a name for the plant, thus placing himself in the number of those who leave plants nameless.
21 See *HN* 21.26 (*Chrysocome*); *HN* 30.16 (*Trixallis* = locust-like insect); *HN* 13.48 (*Phycos*); and many more.
22 *HN* 27.14.

23 Gavin Hardy and Laurence Totelin, *Ancient Botany* (New York: Routledge, 2016), 93, 103–4. Hardy and Totelin discuss the ancient practice of pseudonymity in plant naming, as opposed to synonymity (names in different languages that designate the same plant), homonymity (one name referring to different species), and even anonymity (no known names for a plant, especially if wild and uncultivated). All three practices were surprisingly commonplace.

24 *HN* 25.1.

25 It is also worth pointing out here that by appropriating the *cognomen* into plant nomenclature, the Romans were inviting the comparison of humans with plants and the anthropomorphization of plants. They were also strengthening the idea that discovery (in the sense of an explorer who encounters land and takes credit for knowing about it, despite the presence of indigenous peoples who have occupied it for generations) was a necessary prerequisite for naming.

26 Oscar E. Nybakken, *Greek and Latin in Scientific Terminology* (Ames, IA: Iowa State College Press, 1959), 27.

27 *HN* 25.6. "The most shameful reason for this scarcity is that those who have knowledge of it do not want to teach it, as though what they might hand over to others would then disappear for themselves."

28 Apart from Theophrastus, whose works were unillustrated.

29 C. Singer, "The Herbal in Antiquity and Its Transmission to Later Ages," *Journal of Hellenic Studies* 47 (1927): 5, 7. Crateuas, a botanist who lived under Mithridates of Pontus (120–63 BCE), for whom he was physician, was the author of the *Rhizotomikon*, considered the earliest illustrated herbal. Eleven of Crateuas' illustrations have been preserved in three separate codices of Dioscorides, one of which is at the Morgan Library, the Cheltenham codex.

30 *HN* 25.8. Garrison, "Herbals and Bestiaries," 893.

31 Rita Copeland, "Naming, Knowing, and the Object of Language in Alexander Neckam's Grammar Curriculum," *The Journal of Medieval Latin* 20 (2010): 48. Copeland picks up on the links between names, the things to which they refer, and pedagogy in an analysis of the medieval text *De nominibus utensilium* by Alexander Neckam, which uses short narrative passages to introduce students to a variety of everyday terms. By linking names to actual things, they become solidified. The transmission of knowledge of plants, as Pliny discussed it, is not entirely analogous to this pedagogy but the methodology is quite similar: Neckam solidifies and stabilizes the Latin terms by putting them into a context with other, similar words that deal with the same general subject matter. Likewise, Pliny's point was that by using plants and having experience with them, an individual has a different relationship with that plant's name than if they simply "discover" it, put their name on it, or read about it.

32 Greek is far more synthetic than Latin, allowing for longer, compound words that are more descriptive than a single Latin word would be. Pliny was certainly aware of

this, which means that he saw the content of these Greek names as the problem—that is, despite working with a language that has this incredible communicative feature, the Greeks did not take full advantage of it.

33 All the better if one has access to a garden, as Pliny himself did, to the gardens of Antonius Castor.

34 *HN* 26.5.8.This name is now so obsolete that it is nearly *hapax legomenon* (stated one time only).

35 Perhaps to counteract those things that Nature also produces to our detriment, such as diseases.

36 He also asserted that the Earth, the mother of herbs, produced some of them solely for medicinal purposes.

37 *Papyri Graecae Magicae; die griechischen Zauberpapyri*, ed. K. Preisendanz (Stuttgart: B. G. Teubner, 1928), 12.401–17. English translation in Hans D. Betz and J. Scarborough, *The Greek Magical Papyri in Translation, Including the Demotic Spells* (Chicago, IL: University of Chicago Press, 1992), 167–9. The list includes many other pseudonyms for plants and animal products, including Semen of Helios (white hellebore, 12.432) and Kronos' spice (piglet's milk, 12.430).

38 Translated into Latin under the title *De Succedaneis*, the text is printed in *Claudii Galeni Opera Omnia*, ed. C. G. Kühn (Leipzig: C. Cnoblochii, 1830), 19: 721–47. An adapted version is also in *Paulus Aegineta, Corpus Medicorum Graecorum, IX*, ed. I. L. Heiberg (Leipzig: B. G. Teubner, 1924), vol. II, 401–8; and in Dioscorides, *Materia Medica*, ed. M. Wellmann, 3 vols. (Berlin: Weidmann, 1907), 14.

39 The alignment of herbs with the field of medicine, which most aristocratic Romans saw as a Greek profession that was beneath them, likely contributed to this refusal to learn about plants.

40 See also *HN* 25.32, regarding the plant *Centaurius triorchis*: "There is a third kind of *Centauris*, with the cognomen *triorchis*." Or *HN* 25.68: "There is another *Cyclamen* with the cognomen *cissanthemos*."

41 Pliny did not elaborate on this particular point, but it is worth wondering if any of the people to whom he was referring would have kept medicinal knowledge to themselves. This seems antithetical to the concept of medicine to the modern mind, but if the individuals were not medics, then it is possible that there was no impulse to share what they knew with the broader public.

42 Pliny did not provide a detailed explanation for the selfishness of which he accused Romans in this passage. However, at *HN* 29.8, in addition to claiming that "the height of impunity for having killed a man belongs to the physician alone," he criticized physicians for their greed, calling them a crowd of hoodlums (*multitudo grassantium*), noting that "it is not even shame but their rivals that limit their fees." Pliny saw a troubling pattern in Rome of a transactional transmission of knowledge. Charitable and philanthropic medical institutions were not commonplace in

first-century CE Rome. The earliest such hospitals were Christian and did not begin appearing until around the fourth century (Nutton, *Ancient Medicine*, 314–16). Prior to that, the public hospitals in Rome served a more utilitarian and economic function: they provided medical care to slaves and the indigent because it made sense to keep them healthy enough to work. The only solid evidence we have for public doctors (*iatroi*) is not only heavily debated, but Greek, not Roman. There are exceptions, of course. For example, a physician named Paccius Antiochus is said to have written a letter to the emperor Tiberius when he was on his deathbed *c.* 30 CE. In it, he described a painkiller, which Tiberius then arranged to be available in public libraries, so that anyone could read about it. This remedy is preserved in Scribonius' *Drug Recipes* (#97), but was also attributed to Philonedes of Catania.

43 Nutton, *Ancient Medicine*, 168. In this case, Pliny's claim corresponds to historical and material evidence about physicians in Rome. Based on inscriptions (a biased lot, since most came from imperial households), before 100 CE, only about 10 percent of doctors were Roman citizens; more than 75 percent were slaves or ex-slaves; and less than 5 percent do not have a Greek name.

44 Celsus preserved the famed Mithridates antidote. See *De Re Medica* V.23 for his version of the recipe.

45 Asclepiades of Bythinia, *c.* 129 or 124–40 BCE. It is said that he taught Greek medicine at Rome, a detail that invites skepticism and speculation about how Pliny characterized the translation and dissemination of his works. Brunfels (1530, 272–3) retold this story of Pliny's.

46 Asclepiades' *Praecepta* are no longer extant. There are some *Precepts* preserved under the title *Parangeliai*, sometimes attributed to Asclepiades, although they are generally treated as part of the Hippocratic corpus. Based on Pliny's and Galen's takes on Asclepiades, it is unlikely that these *Precepts* are those of Asclepiades, since the author says that in the practice of medicine, one should not focus on theories, but on a combination of experience and deductive reasoning.

47 Cicero, *De Optimo Genere Oratorum* (DOGO) 5.

48 Whether Asclepiades' *Precepts* were considered technical or philosophical is unknown. The *Parangeliae* sometimes attributed to Asclepiades reads as more philosophical than technical, but is also quite unlikely to actually be Asclepiades' *Precepts*.

49 Silvia Rizzo, "A Study of Words Is also by Definition a Study of Things," *Il Lessico filologico degli umanisti* (Roma: Edizioni di Storia e Letteratura, 1973), x.

50 *HN* 26.6.10.

51 *HN* 26.6.

52 That is, when they are blooming, identifiable, etc.

53 Again, this may simply be a dig at the Greeks, who were often seen by the Romans as soft and effeminate and evidently neither suited nor inclined to physical exertion and discomfort.

54 Colin Webster, “Heuristic Medicine: The Methodists and Metalepsis,” *Isis* 106 (2015): 665. Webster analyzes the Methodist sect of physicians, who rejected both the Hippocratic rationalist approach to medical diagnoses and the empiricist approach, which were focused on etiology, in favor of three simplified, non-causal “manifest commonalities.” Asclepiades in particular used the rhetorical method of metalepsis, in which metonymy is used in inappropriate places, as when he defines “plague” as “an unusual condition of living creatures in the particularly affected locality, because of which [condition] they are prone to be visited with deadly diseases arising from a common cause,” per Caelius Aurelianus. Basically, Asclepiades defines plague as “an unusual condition” because of which people are stricken with disease (also plague), which have a common cause (plague).

55 Pliny’s summary (*HN* 26.7.13) of their reception in Rome could be taken as tongue-in-cheek. Despite realizing that they could have come up with such ideas on their own, Asclepiades was applauded and won over nearly the whole of humanity, “just as if he had arrived because he was sent from heaven.”

56 *HN* 26.8.14.

57 Herophilus of Chalcedon (335–280 BCE), who along with Erasistratus of Ceos (*c.* 304–*c.* 250 BCE) founded a school of anatomy in Alexandria.

58 The gender and number of the appellation, *frigida danda,* is odd, since it is neuter accusative plural, and thus literally means “cold things that must be administered.” Technically, Asclepiades’ preferred *cognomen* is not active, signifying the one administering remedies, but passive, signifying the remedies themselves. In the medical manuscript Anonymus Londinensis XXIV 30, he is called “Asclepiades the Wine Giver.”

59 Pliny tells of an incident whereby Asclepiades is supposed to have saved a man’s life by plucking him from his own funeral pyre.

60 Galen had evidently discussed Asclepiades at length in a now lost work, to which he refers at *ONF* 1.xii.

61 *ONF* 1.xiii.

62 Namely, the truth in question is that drugs are not the cause of the bodily fluids that are evacuated when they are administered.

63 *ONF* 1.xiii.41.

64 *ONF* 1.xiv.51.

65 Ibid.

66 Ibid.

67 In a medicinal context, this would include the diagnosis of a patient.

68 *HP* 1.I.

69 In the literal sense of production or generation from seed.

70 *HP* 2.I.1.

71 *HP* 2.IV.3.

72 See *HN* 36.29, where Theophrastus and Mucianus are cited for the idea that stones can generate other stones.

73 Bradley C. Bennett, "Doctrine of Signatures: An Explanation of Medicinal Plant Discovery or Dissemination of Knowledge?," *Economic Botany* 61 (2007): 246–55. This is an example of the doctrine of signatures, which states that the curative powers of a plant, or any natural object, can be discovered by reference to the body part it resembles. Bennett traces its historical origins and locates it in both Pliny and Dioscorides. For example: "The herb *Scorpius* resembles the tail of the scorpion and is good against his biting" (Dioscorides 2000). In the case of Pliny, Bennett does not provide an actual citation, but rather references the 2006 *Encyclopedia Britannica* entry on the doctrine of signatures. This is because, in fact, Pliny did not explicitly aver the doctrine, though there are numerous instances in Books 23–7 where he referenced or hinted at it. For example, *HN* 23.16, regarding the herb *bryony* (*Bryonia dioica*), which Pliny noted is a white vine. One of its medicinal uses, once mixed with water, is to extract splintered bones, for which reason it was often called "white bryony." Bennett argues that the doctrine of signatures was a mnemonic for disseminating knowledge in pre-literate and pre-scientific societies.

74 The Hippocratic Oath no doubt plays a role in Pliny's take on abortifacients. "I will neither give to anyone a lethal drug, if I am asked, nor will I instruct anyone in such a plan; and similarly, I will not give a woman an abortive pessary." One of Pliny's biggest criticisms of Asclepiades involved his *magicae vanitates* (magical deceptions). Pliny abhorred those who exploited others by using magic, but at the same time, when discussing so-called magical herbs, he said that he was forced by accounts of their healing properties to admit that there is some truth in them (*coguntque confiteri multum esse veri quod supersit*). See *HN* 25.5.

75 For example, Charmis of Massilia, who rejected the previous trend of using hot baths, and instead convinced his patients to take cold baths, even in the dead of winter. *HN* 29.5.

76 *HN* 29.1.

77 For instance, Celsus' *De Medicina* was likely written before the *Historia Naturalis.*

2 Otto Brunfels' Reception of Pliny

1 Otto Brunfels, *Herbarum Vivae Eicones* (Strassburg: Johann Schott, 1530–2).

2 This can be deduced from the full title: "Living Images of Herbs, in Imitation of Nature, with the Highest Diligence and Skill in the Portrayal, together with the Effects of the Same, in Gratitude for that Ancient [Authority], and Medicines from the now Reborn [profession of the] Herbalist."

3 T. A. Sprague, "The Herbal of Otto Brunfels," *Journal of the Linnean Society of London, Botany* 48 (1928): 79.

4 Sprague, "The Herbal of Otto Brunfels," 80; cf. Leonhart Fuchs, *De Historia Stirpium* (Basil: In officina Isingriniana, 1542), 18. There are some errors in Brunfels, in which some illustrations do not match the plant description they accompany. For example: *Herba Roperti* (*Germanium Robertianum L.*), which is actually *Erodium cicutarium L.*

5 Agnes Arber, *Herbals, their Origin and Evolution: A Chapter in the History of Botany, 1470–1670* (Cambridge: Cambridge University Press, 1912 [1986]), 324.

6 Brunfels cannot be credited with creating a new language or linguistic phenomenon. Even modern botanical Latin is not really a new language, but a strict set of rules and suggestions for using an existing language with which many botanists are unfamiliar but required to have a basic knowledge. By "formal," I am therefore referring to the emergence in the twentieth century of rules and recommendations for nomenclature and the description of plants.

7 The *editio princeps* was printed in Venice in 1469 by Johannes de Spira and the 1470 edition was printed in Rome by Sweynheym and Pannartz and edited by the Bishop of Aleria, GiannAndrea Bussi.

8 Niccolò Leoniceno, *De Plinii et Aliorum in Medicina Erroribus* (Ferrara: Rossi, L. de Grassis, A., 1492).

9 The transliteration of the Greek tends to switch back and forth between *cissos/cisthon* and *kissos/kisthon* in the sources. Outside of direct quotations, I will use *cissos* and *cisthon*, which correspond to a larger number of Latin sources.

10 Pliny, *Historia Naturalis* 16.145 and 24.81. See Leoniceno, *De Plinii*, 1–2.

11 Ermolao Barbaro, *Hermolai Barbari Castigationes Plinianae et in Pomponium Melam / Edidit Giovanni Pozzi* (Passau: In aedibus Antenoreis, 1493), aii verso.

12 Ermolao Barbaro, *In Dioscoridem Corollariorum Libri Quinque* (Cologne: Ioan. Soterem, 1530).

13 Ermolao Barbaro, *Historia Naturalis*, ed. Ermolao Barbaro, rev. Johannes Baptista Palmarius (Venice: Bernardinus Benalius, 1497).

14 He refers to the author of this text as "Pandectarius."

15 Brunfels is referring to the *Pandectarum Medicinae*, or *Pandectae Medicinae*, published *c.* 1317 by Matthaeus Silvaticus. Silvaticus used as a partial reference for his encyclopedia the *Synonyma Medicinae* of Simon Ianuensis, which pre-dates the *Pandectarum* by a few decades and is similar to a dictionary listing medicinal preparations.

16 Brunfels, *Herbarum Vivae Eicones*, 8.

17 Brunfels, *Herbarum Vivae Eicones*, fol. a3r. The closeness of the connection between Barbaro and Pliny pre-dates Brunfels. The 1513 edition of Pliny, printed by Melchior Sessa in Venice, is the first illustrated edition and is bound together with Barbaro's

Castigationes. The illustrations do not depict natural historical phenomena, whether flora, fauna, or mineral, in the text, but are entirely decorative. It is also worth pointing out that Barbaro was neither a botanist nor a physician. His expertise is with regard to the recension and transmission of Pliny's text: he uses *recensio ope ingenii sui* (recension via skill) and *recensio ope codicum* (recension via manuscripts) both.

18 Brunfels, *Herbarum Vivae Eicones*, 17.

19 London, British Library, Egerton MS 747, fol. 13v.

20 Also known as the *Latin Herbarius*, *Herbarius in Latino*, *Aggregator de Simplicibus*, *Herbarius Patavinus* (Herbarius Maguntie impressus. Peter Schöffer. Mainz), fol. xxvr. As in Brunfels, the German names are also provided, in this case, *Bethonich*.

21 It is important to note that the author of this herbal does not cite any of the standard medieval herbalists for this particular entry, although they do for other herbs.

22 Sorcha Rizzo, "A Study of Words Is also by Definition a Study of Things," *Il Lessico filologico degli umanisti* (Roma: Edizioni di Storia e Letteratura, 1973), ix.

23 Ibid., x.

24 Leonhard Fuchs, *De Historia Stirpium*, fol. 3v.

25 Brunfels, *Herbarum Vivae Eicones*, 1.

26 Ibid.

27 On page 135, Brunfels described his sources for the plant *Narcissus*, Theophrastus and Dioscorides, as men who have themselves learned from experience rather than from books.

28 Brunfels indulged in the somewhat novel practice of providing an index of his ancient and contemporary sources, helpfully titled *Autores Quorum Testimoniis in Hoc Opere Usu Sumus*.

29 Brunfels, *Herbarum Vivae Eicones*, 8; Pliny *HN* 24.1.4.

30 Brunfels, *Herbarum Vivae Eicones*, 8.

31 Ibid., 17.

32 Ibid., 17–18.

33 The other is the printing history of Pliny.

34 Brunfels, *Herbarum Vivae Eicones*, 18.

35 Barbaro, *In Dioscoridem Corollariorum*, 42.

36 Leoniceno, *De Plinii*, 30 used this plant as part of his critique of Pliny, claiming that he identified two types of *Arction*, but that the one Pliny called *Echion* or *Personata* was in fact what is now called either *Bardana* or *Lappa maior*. Pandulfo Collenuccio, *Pliniana Defensio Adversus Nicolai Leoniceni Accusationem* (Ferrera: Belfort, André, 1493), 68, disagreed, arguing that Pliny did not equate *Echion* and *Arction*.

37 Brunfels, *Herbarum Vivae Eicones*, 45.

38 Ibid., 89.

39 Lucie Čermáková and Jana Černá, "Naked in the Old and the New World: Differences and Analogies in Descriptions of European and American *herbae nudae* in the Sixteenth Century," *Journal of the History of Biology* 51 (2018): 69–106.
40 Brian Ogilvie, *The Science of Describing: Natural History in Renaissance Europe* (Chicago, IL: University of Chicago Press, 2006), 139.
41 The author of the *Botanologicon* (1534).
42 Čermáková and Černá, "Naked in the Old and the New World," 71.
43 Brunfels, *Herbarum Vivae Eicones*, 217.
44 Brunfels' use of the adjective *nudus* can be contrasted with Ermolao Barbaro, who explained how for some of the errors in Pliny, he emended with the "barest of expressions," whereas for others, he had much more to say (Barbaro, *Castigationes*, aiir).
45 Brunfels, *Herbarum Vivae Eicones*, 95. I refer to this herb as *semi-nuda* because Brunfels was aware of its Latin and Greek names. The issue was that he could not figure out which of them corresponded to the plant he was describing, or to the three German names of which he was aware.
46 A genus in the *Asteraceae* family, i.e., goldenrods.
47 Brunfels, *Herbarum Vivae Eicones*, 217.
48 Ibid., 218.
49 Ibid.
50 Ibid.
51 Čermáková and Černá, "Naked in the Old and the New World," 71.
52 Paula Findlen, *Possessing Nature: Museums, Collecting, and Scientific Culture in Early Modern Italy* (Berkeley, CA: University of California Press, 1994), 156. Cf. Čermáková and Černá, "Naked in the Old and the New World," 71.
53 Many botanists after Brunfels took the time to devise Latin names for the new plants they encountered, although, with the modern scientific definitions of genus and species not yet claimed and with the rules of nomenclature not yet established, authors frequently resorted to simply naming a "new" species as the "first," "second," or "third" of a particular genus. By 1601, this practice had led to a large number of novel plant names that could be traced back to specific sixteenth-century authors. In Giovanni Pona's *Plantae seu Simpliciae* (1608), one finds such plant names as *Echium fuchsii*, *Rhamnus tertius Matthioli*, *Scorpioides Matthioli*, etc. Notably, Pona did not list any plants named after Brunfels, an indication that he got his wish.
54 Brunfels, *Herbarum Vivae Eicones*, 68.
55 Ibid.
56 Ibid., 183.
57 Ibid.
58 Presumably this was done in person, though this individual could also have been using an illustration.

59 Brunfels provided illustrations of each of the two species, masculine and feminine, but no description of the German plant. The name is not helpful, since *Durchwachß* is related to *durchwachsen*, "to grow through." Thus, what Brunfels is saying is that the first plant shown to him by herbalists, which they called *Perfoliata*, did not look anything like what he was expecting, namely, *Durchwachß*. The second plant, however, did and the two illustrations are almost certainly of this alternative *Perfoliata*. There are gaps in the narrative, but it appears that Brunfels ascertained somehow that the plant *Durchwachß* was called *Perfoliata* and in an attempt to verify this by obtaining a plant by that name from the herbalists, he found himself with two different plants.

60 Brunfels, *Herbarum Vivae Eicones*, 184.

61 I take it that *plura* (more) in this statement is modifying a missing term such as *genera*, although the absent object of the verb could also just be "things," as in "information."

62 William Stearn, *Botanical Latin: History, Grammar, Syntax, Terminology and Vocabulary*, 4th edition (Portland, OR: Timber Press, 1995), 194.

63 Helen Choate, "The Earliest Glossary of Botanical Terms; Fuchs 1542," *Torreya* 17 (1917): 186–201.

64 Fuchs, *De Historia*, B3 has the original Latin definition: *Herba est quod a radice foliatum, sine caudice provenit, saepiusque semen in caule fert.*

65 Fuchs, *De Historia*, 631.

66 Brunfels, *Herbarum Vivae Eicones*, 184.

67 Ibid. Brunfels' Latin is convoluted, but it seems that he considered his task to be provocative, not pedagogical in such situations. If he were concerned with teaching others about *Narcissus*, he would have left it out, since he could not make sense of it.

68 Fuchs, *De Historia*, a3v–a4r stated in his preface that the entirety of the profession of medical herbalism had fallen into the hands of pharmacists, fools, and those who are superstitious, i.e., country folk, old women, and uneducated men.

69 Brunfels, *Herbarum Vivae Eicones*, 68. See the entry for *Serpentaria* on page 62.

70 Ibid.

71 Ibid., 82.

72 Ibid., 86. Compare this to his treatment of the herb *Tormentilla*. Brunfels did everything he could to add external authority to his take on the problem, refusing to state a firm opinion and saying, *nihil habeo quod respondeam* (I have no response to make). Instead, he cited Pliny 25.9, who in turn got his information from Theophrastus. They both described the plant as *Pentaphyllon*, in such agreement that they must have been accurate enough to serve as models for a painter, an indication that Brunfels saw just as much value in textual descriptions for identifying herbs as in illustrations.

73 Brunfels, *Herbarum Vivae Eicones*, 114; cf. Diosc. 2.129.

74 Pliny, *HN* 27.39
75 Barbaro, *Castigationes*, fol. 200r.
76 Brunfels, *Herbarum Vivae Eicones*, 117. The original is in Barbaro, *Castigationes*, fol. 61v. Columella has a recipe for salad that consists of just such garden herbs, *On Agriculture* XII.59.
77 Brunfels, *Herbarum Vivae Eicones*, 117.
78 Ibid., 133.
79 It has two German names, one for the first flowering in March, and another for the September flowering.
80 Brunfels, *Herbarum Vivae Eicones*, 133.
81 Ibid.
82 Ibid.
83 Ibid., 145.
84 Ibid., 146.
85 This wonder is also unlike that expressed by the Europeans in the New World: Brunfels was dismayed in such a manner that he was approaching apostasy, whereas the New World naturalists were filled with awe at the unfamiliar productions of Nature.
86 Regarding *Cynoglossa*, he had not yet verified the second type described by Pliny, which has small burs (*lappas minutas fert*). This is in reference to Pliny's statement that in addition to the already known *Cynoglossa* there is another similar one, which has tiny burs (25.41). Regarding *Hepatica*, he had only discovered in the accepted authors what he himself recalls about it. See Brunfels, *Herbarum Vivae Eicones*, 177 and 192, respectively.
87 Brunfels, *Herbarum Vivae Eicones*, 133.
88 Attributed to Matthaeus [Silvaticus] and Matthaeus Moretus, and published in Venice.
89 The text is in the original Latin, but the front matter and primary names are all in German.
90 Serapion of Alexandria, whose medical works are not extant and whose opinion is known by way of Galen, Dioscorides, Celsus, etc. In this particular instance, the attribution to Serapion is by way of Dioscorides.
91 *Liber Pandectarum Medicinae* 1488: fol. 42r.
92 Brunfels, *Herbarum Vivae Eicones*, 68.
93 Ibid., 221–2.
94 A. G. Morton, *History of Botanical Science: An Account of the Development of Botany from Ancient Times to the Present Day* (London: Academic Press, 1981), 118.
95 International Association for Plant Taxonomy. *Twentieth International Botanical Congress*. "International Code of Nomenclature for Algae, Fungi, and Plants (Madrid Code). Adopted by the Twentieth International Botanical Congress, Madrid, Spain,

July 2024" (Chicago, IL: University of Chicago Press, 2025). The influence of Linnaeus' work can be seen in the numerous rules and recommendations that cite his system. See, for example, Article 20.2, which states that a species name should be "in accordance with the binary system of Linnaeus."

96 Kaspar Bauhin, *Pinax Theatri Botanici* (Basil: Sumptibus & typis Ludovici Regis, 1623), 17. Bauhin quotes Galen, who referred to *materia medica* as "innate qualities" (*bonae indoles*) of plants.

97 Barbaro, *Castigationes*, aiir.

98 William Stearn, *Botanical Latin*, 17. Stearn argues that this approach to plants is first seen in Theophrastus.

99 There is a relationship between this anxiety and distrust on Brunfels' part, and the influence of alchemical and alchemy-adjacent treatises on Italian scholarship in the fifteenth century. This can be seen in such texts as Nikolaus Cusanus' 1440 *De Docta Ignorantia*, which argues that scientific inquiry and rationality are insufficient to understand God; that speculation, which is called *ignorantia*, was a crucial component of true knowledge.

3 Botany from Leonhard Fuchs to Linnaeus and Beyond

1 Hieronymus Bock (1498–1554) published his *KreüterBuch* in German in 1539 without illustrations, though later editions did include them, some recycled from Fuchs. The first Latin edition was printed in 1552. He included *Gauchblům*, but his focus was primarily on German-language herbals, not Latin.

2 Antonio Musa Brasavola, *Examen Omnium Simplicium Medicamentorum* (Lyon: J. et F. Frellaeos, 1536) is a compendium of medicinal plants, simples being one-ingredient concoctions that are generally made from a herb. Similar texts had been proliferating since the advent of printing. Brasavola's text matches vernacular names of plants native to Italy to ancient names, especially in Dioscorides. It is entirely possible to argue, therefore, that Brasavola's text is on equal footing with Brunfels' in the development of botany and its formal language. However, it was Brunfels, a fellow German, to whom Fuchs primarily responded and reacted. Another issue concerns stylistics. Brasavola's text is a dialogue between himself and an elderly man named Senex, whereas both Brunfels and Fuchs retained the catalog format of the herbal tradition. Brasavola's writing about plants is not as stylistically connected to the increasingly sparse and formal style of botanical Latin.

3 Other editions in 1754, 1767, and 1791 attest to the immediate popularity of the text.

4 Admittedly, this is a simplification. Many plants have been recategorized and renamed since Linnaeus' time. Almost every issue of the journal *Taxon* includes proposals for renaming or reclassifying some genus or species, and the International

Botanical Congress regularly handles such issues via its Committee on Nomenclature.

5 Proposals to reclassify or rename taxa that have already been validly published are not the focus of this statement, and more importantly, they are not written in Latin.

6 In fact, Fuchs employed three different woodcut artists and had them combine images of certain plants at different stages of their life cycle into one, hyper-realistic image. An illustration of a flowering plant might also include withering leaves, etc. This resolves the issue of space in a printed text, eliminating the need for multiple images of one plant, but reintroduces the ancient problem of identification in the wild. See Chapter 5 for further discussion of the efficacy of illustrations.

7 Fuchs himself has been heavily criticized by historians of botany and his reputation as a rigorous scientist is only slightly better than Brunfels.'

8 Fuchs also was unable to match them to Greek and Latin plant names and had little or nothing to add or alter in Brunfels' descriptions of them.

9 Fuchs had previously made his name by publishing in 1530 a book about the errors made by other physicians, entitled *Errata recentiorum medicorum.* It was not Brunfels he had in mind; Fuchs was critical of the Arab tradition and the ways in which Greek medicine had been misunderstood and misrepresented. He also disapproved of Pliny's misinterpretations of the Greek medical texts and it may be assumed that he was also critical of Brunfels' adoration of Pliny.

10 Leonhard Fuchs, *De Historia Stirpium* (Basil: In officina Isingriniana, 1542), fol. 5r.

11 Ibid., fol. 5r.

12 Fuchs, *De Historia Stirpium*, 14–15.

13 Fuchs, *De Historia Stirpium*, a2r.

14 Brunfels, *Herbarum Vivae Eicones*, 68.

15 Fuchs says that Pliny called it *Lepidion*, but the term cannot be found in modern editions of Pliny.

16 Fuchs, *De Historia Stirpium*, 324.

17 Ibid.

18 In his brief description of *Gauchblům*, Brunfels, *Herbarum Vivae Eicones,* 218 used *velint* and *pellat*, both subjunctives.

19 Under *Vires*, he cited Pliny for the fact that when formed into a paste, it is very useful for the hips and other joints in conjunction with a hot bath (Fuchs, *De Historia Stirpium*, 326).

20 He borrowed heavily from Dioscorides, Pliny, and Pandectarius in this entry. However, this is not the same plant as *Gauchblům.* It is important to remember that *Nasturtium* is technically a generic name, and there are many species of plant that fall under its classification.

21 Fuchs, *De Historia Stirpium*, 324.

22 Ibid., 326. This remedy is also found in Galen.
23 Another of the plants for which Brunfels provided an *iudicium*.
24 Fuchs, *De Historia Stirpium*, 671–2.
25 Andrea Cesalpino, *De Plantis Libri xvi* (Florence: Georgius Marescottus, 1583), Book 8, cap. 70.
26 Cesalpino, *De Plantis Libri*, 318.
27 Cesalpino, *De Plantis Libri*, 363.
28 His French name is often used instead: Charles de L'Ecluse.
29 Carolus Clusius, *Caroli Clusii Atrebatis Rariorum Plantarum Historia, Quae Acceserint Proxima Pagina Docebit* (Antwerp: Ex Officina Plantiniana, Johann Moretus, 1601), 13. The irony being that a Spanish name for the tree would have derived obliquely from Latin.
30 Kaspar Bauhin, *Pinax Theatri Botanici* (Basil: Sumptibus & typis Ludovici Regis, 1623), 17.
31 Ibid.
32 Bauhin, *Pinax*, 569.
33 Bauhin, *Pinax*, 103.
34 *De Materia Medica* 2.c.185
35 *Historia Plantarum* 7.1
36 *HN* 1.20.13.
37 Barbaro had employed a similarly sparse technique. He simply listed the errors in Pliny's *Historia Naturalis*, some 5,000 of them, in order, and corrected them.
38 *Nasturtium pratense magno flore* is *Flos cuculi*; and *Nasturtium sylvestre Osyridis folio* is *Bursa pastoris minor*.
39 Bauhin, *Pinax*, 103.
40 Bauhin, *Pinax*, 104. Pietro Andrea Gregorio Matthioli (1501–*c*. 1577), a medical botanist and author of the *Discorsi* (1554) on Dioscorides' *De Materia Medica*.
41 Bauhin, *Pinax*, 104.
42 *N. hortense latifolium*, *N. hortense vulgatum*, and *N. hortense crispum* are all now known as *Lepidium sativum* L. Meanwhile, *Nasturtium pratense magno flore* is *Cardamine pratensis,* which is how Linnaeus reclassified *Nasturtium agreste*.
43 Bauhin, *Pinax*, 177
44 *HN* 1.21.c.11 and *HP* 6.7.
45 The *differential* diagnosis traditionally stated the ways in which a plant differed from others in the same group, in as few words as possible, with no inessential information. The *essential* diagnosis is similar, but expresses the idiosyncrasies of a plant, which are therefore essential to it and are not shared by other plants in the same genus or family. See William Stearn, *Botanical Latin: History, Grammar, Syntax, Terminology and Vocabulary*, 4th edition (Portland, OR: Timber Press, 1995), 143–4 for a summary. John Lindley, *The Elements of Botany, Structural, Physiological, &*

Medical (London: Taylor and Walton, 1841) provides an early explanation of the differences between differential and essential diagnoses.

46 Carl Linnaeus, *Species Plantarum* (Stockholm: Laurentius Salvius, 1753), 374.

47 The effects of re-Latinizing names in Romance languages, which themselves derive from Latin, are an interesting study in and of themselves. Consider, for example, Avenio, the accepted botanical geographical name for Avignon, or Castella Nova for Castilla la Nueva in Spain.

48 Stearn, *Botanical Latin*, 144.

49 Stearn, *Botanical Latin*, 154.

50 Carl Linnaeus, *Philosophia Botanica* (Stockholm: Godofr. Kiesewetter, 1751), no. 328.

51 Alphonse De Candolle, *La Phytographie, ou l'Art de Décrire les Végétaux* (Paris: G. Masson, 1880), 166 (Article IV). Cf. Stearn, *Botanical Latin*, 154.

52 Stearn, *Botanical Latin*, 154.

53 Stearn, *Botanical Latin*, 154–5.

54 Stearn, *Botanical Latin*, 194.

55 Linnaeus, *Species Plantarum*, (2): 915 (or: 656).

56 Ibid. (dwarf, cf. Bocc. mus. 2.p.171.t.166.)

57 Linnaeus, *Species Plantarum*, 914.

58 K. Marhold, "Typification of the Linnaean Names of the Genus *Cardamine* (Cruciferae)," *Botanical Journal of the Linnean Society* 121 (1996): 124 has identified four possible types, or specimens, to support Linnaeus' reclassification, one in the Linnaean herbarium, London (LINN no. 835.15), identified as the lectotype of *Cardamine pratensis* (a specimen chosen when the original author cited more than one). The other three specimens are in the Clifford herbarium (BM), that is, the Hortus Cliffortianus.

59 Carl Linnaeus, *Genera Plantarum* (Paris: M. A. David, 1743), #645.

60 Linnaeus, *Genera Plantarum*, 247.

61 Linnaeus, *Genera Plantarum*, xxxi–xxxii.

62 See William Botting Hemsley, "The History of the Botanical Magazine," in *A New and Complete Index to the Botanical Magazine* (London: Lovell Reeve, 1906), xxiv.

63 Henry Cranke Andrews, *The Botanist's Repository for New and Rare Plants* (London: The Author, 1799–1828), vol. 2 (1800), plate 126.

64 *Hibbertia* is feminine but several species have masculine names, such as *H. diamesogenos* and *H. lasiopus,* indicating a degree of confusion about the gender of the term in botanical nomenclature. See Chapter 5 for a discussion of the naming of this genus.

65 Edmond Boissier, *Elenchus Plantarum Novarum Minusque Cognitarum quas in Itinere Hispanico* (Geneva: Typographia Lador et Ramboz, 1838), 83. The generic name *Quercus* is the classical Latin term for "oak tree." The synonym *Quercus faginea* Lam. (Gall Oak) is now the preferred, valid name. However, at the time Boissier was

writing, he believed this to be an entirely different species of *Quercus* than the one described by Lamarck in 1785. See Francisco M. Vázquez, María Gutiérrez, and Soledad Ramos, "Diversity of the Gall Oak (*Quercus faginea* Lam.) in the Iberian Peninsula," *International Oaks: The Journal of the International Oak Society* 21 (2010): 121–36.

66 The Latin here is concise and best viewed all together: *foliis oblongo-lanceolatis pallide virentibus deciduis coriaceis supra glabris subtus plus minusve tomentosis reticulato-venosis, margine undulatis, distanter grosse et irregulariter dentatis, dentibus latis acutis.*

67 Boissier, *Elenchus Plantarum*, 83.

68 Stearn, *Botanical Latin*, 146.

4 Botanical Latin and Nomenclature in the Twentieth Century

1 It goes without saying that colonization was worldwide and not restricted to the Americas. The decision to focus the geographical discussion on the Americas is based only on the need to provide a robust counterexample to the study of plants in Europe in roughly the same time period. The same arguments can be made about the effects of botanical colonization elsewhere.

2 Evan T. Jones, "The Matthew of Bristol and the financiers of John Cabot's 1497 voyage to North America," *English Historical Review* 121 (2006): 778–95. The earliest known expedition to North America was led by the Venetian John Cabot (*Giovanni Caboto*, *c.* 1450–99), who in 1496 was issued a *Letter patent* by King Henry VII, granting him a monopoly on trade routes to China as well as any new lands, provided he set sail from Bristol. He arrived in either modern-day Newfoundland or Cape Breton Island in 1497. William Weston (*c.* 1445–*c.* 1504), a merchant from Bristol, may have been part of Cabot's expedition, but made his own voyage in either 1499 or 1500, when he received a reward from Henry VII of £30 for his troubles. Almost nothing definitive is known about Weston's voyage, but it is discussed in a "Letter from King Henry VII to his Lord Chancellor regarding Weston's planned expedition to 'the new founde land'": The National Archives, C82/332 piece 61, out of 74.

3 Suriname excepted, having only gained their independence from the Dutch Republic in 1975.

4 Anecdote told in Cristoph Irmscher, *The Poetics of Natural History* (New Brunswick, NJ: Rutgers University Press, 2019), 5. Source: "Extract of a Letter, Written by John Winthrop (1670)," in *The Puritans: A Sourcebook of their Writings*, ed. Perry Miller and Thomas H. Johnson, revised edition (New York: Harper & Row, 1963), vol. 2, 740.

5 Carl Linnaeus, *A General System of Nature* (Swansea: Printed by Voss and Morris, 1806), vol. I, 2–3.

6 John Gerard, *The Herball, or, Generall Historie of Plantes* (London: John Norton, 1597), 77.

7 H. P. Biggar, *The Voyages of Jacques Cartier*. Publications of the Public Archives of Canada, 11 (Ottawa: F. A. Acland, 1924), 204.

8 Charlotte Erichsen-Brown, *Medicinal and Other Uses of North American Plants: A Historical Survey with Special Reference to the Eastern Indian Tribes* (New York: Dover Publications, 1979), ix.

9 Jonathan Carter, *Travels Through the Interior Parts of North America, In the Years 1766, 1767, and 1768* (London: C. Dilly, 1781), 505.

10 Carl Linnaeus, *Species Plantarum* (Stockholm: Laurentius Salvius, 1753), 536.

11 Ward was not the first to think of such a case, but he was the first to publish his findings, in his 1842 *On the Growth of Plants in Closely Glazed Cases.*

12 Jürgen Leonhardt, *Latin: The Story of a World Language* (Cambridge, MA: The Belknap Press of Harvard University Press, 2013), 130.

13 The Jesuits were expelled from the Portuguese colonies in 1759 by the Marquis of Pombal well before Linnaeus' work became influential enough to be taught by them.

14 J. Ramirez Cabañas, ed., *Bernal Díaz del Castillo, Historia Verdadera de la Conquista de la Nueva España* (México: Pedro Robredo, 1944), 130.

15 There is a fascinating symbiosis here: English had long borrowed its vocabulary from Latin and Greek, and now loaned itself out to botanical Latin.

16 Andrew Laird, *The Epic of America: An Introduction to Rafael Landívar and the Rusticatio Mexicana* (London: Bloomsbury Academic, 2006), 149 = *Rusticatio* IV.28. A couple of lines later at IV.32, Landívar uses the phrase *carnosas . . . frondes* (fleshy leaves) to refer to the cactus' physical characteristics. In his notes, he remarks that he really should call them "branches" but that he has used a Latin word for "leafy branch" in order to comply with usage (*usui*) and accepted way of speaking (*receptoque loquendi modo*). Laird, *The Epic of America*, 268.

17 Laird, *The Epic of America*, 25. De Pauw was hardly the only one to make such arguments. George-Louis Leclerc Buffon, a French naturalist, who influenced De Pauw, had also argued that since America was a geologically immature continent, everything in it, including people, was also immature and therefore inferior to Europeans.

18 Laird, *The Epic of America*, 13.

19 He recorded the baptismal names of three indigenous painters with whom he worked: Antón, Baltazar Elías, and Pedro Vázquez.

20 Nicolás León, *Catálogo Bibliográfico, Biográfico y Crítico de Autores y Escritos Referentes a Vegetales de México y sus Aplicaciones, desde la Conquista hasta el Presente* (Mexico: Oficina Tip. de la Secretaría de Fomento, 1895), 797.

21 The full title has the phrase, *lo que el Doctor Francisco Hernández Escribió en Lengua Latina.* (what Doctor Francisco Hernández wrote in the Latin language). Recchi's corpus was purchased by an Italian physician, Federico Angelo Cesi (1585–1630), founder of the Accademia dei Lincei. Cesi himself published a heavily edited version of Ximénez's printed edition in compendium form in Latin under the title *Rerum medicarum Novae Hispaniae thesaurus* in 1628.

22 Francisco Ximénez, *Quatro Libros de la Naturaleza* (Mexico: en casa de la Viuda de Diego Lopez Daualos, 1615), A2-1.

23 Ximénez, *Quatro Libros*, Libro Segundo, fol.87r.

24 Ximénez, *Quatro Libros*, Libro Primero, fol.34v. In fact, four subspecies or types of cocoa are described, which mainly differ in size, with the largest two both being called *cacahuaquahuitl*, the third, which is smaller, being called *xuchicacahuaquahuitl*, and the fourth, the smallest, being called *tlalcacacahuaquahuitl*.

25 Simon Varey, *Searching for the Secrets of Nature: The Life and Works of Dr. Francisco Hernández* (Stanford, CA: Stanford University Press, 2000).

26 Daniela Bleichmar, "Visual Culture in Eighteenth-Century Natural History" (PhD dissertation, Princeton University, 2004), 2.

27 Part I is called *Historia Medicinal de las Cosas que se Traen de Nuestras Indias Occidentales.* Sevilla: Hernando Diaz. This text was so popular that it was translated into Latin, French, and English (Rohde 1922, 120–41), the latter in 1596 under the title *Joyfull Newes out of the New-found Worlde.*

28 Part II is called *Segunda Parte del Libro des las Cosas que se Traen de Nuestras Indias Occidentales, que Sirven al uso de la Medicina.*

29 A Royal order of 1789 demanded that those in administrative roles in the colonies produce reports on such materials and send specimens back to Spain.

30 Richard Brown, *Knowledge is Power: The Diffusion of Information in Early America, 1700–1865* (New York: Oxford University Press, 1989) provides in-depth analyses of the information-controlling networks operating in North America, such as New England rural clergymen and lawyers in provincial Massachusetts. Brown uses case studies to show how knowledge of plants, crops, medicine, and agriculture granted autonomy and power to those without traditional, classical knowledge of literature, language, law, etc.

31 Bleichmar, "Visual Culture," 3.

32 Maria Sibylla Merian, *Metamorphosis Insectorum Surinamensium. Ofte verandering der Surinaamsche insecten* (Amsterdam: Voor den auteur, als ook by G. Valck, 1705), 45.

33 Henry Wellcome, *Souvenir Cinchona Tercentenary Celebration and Exhibition at the Wellcome Historical Medical Museum*, conserv. L. W. G. Malcolm (London: Wellcome Historical Medical Museum, 1930), 103, no. 985. Attributed to "J. Hawkeens," which may be a misspelling of John Hawkins.

34 Wellcome, *Souvenir Cinchona*, 103, no. 985. From a letter to Joseph Banks, dated February 8, 1794.

35 In fact, the Viceroy's diary was discovered in 1930 and it contradicted Bado's account: Ana de Osorio died a few years before he arrived in Peru with a second wife, who had no known health scares. Historians now credit the Jesuit Barnabé de Cobo (1582–1657) with bringing the bark to Europe from Lima in 1632.

36 Quinine was isolated by Pierre Joseph Pelletier and Joseph Bienaimé Caventou in 1820, most likely from *Cinchona pubescens*.

37 Wendy Makoons Geniusz, *Our Knowledge Is Not Primitive: Decolonizing Botanical Anishinaabe Teachings* (Syracuse, NY: Syracuse University Press, 2009), 17–18.

38 Lucille Brockway, *Science and Colonial Expansion: The Role of the British Royal Botanic Garden* (New Haven, CT: Yale University Press, 2002).

39 Erichsen-Brown, *Medicinal and Other Uses of North American Plants*, 39–46.

40 Bernard Cohn, *Colonialism and Its Forms of Knowledge: The British in India* (Princeton, NJ: Princeton University Press, 1996), 3.

41 Karl Koch, *Wanderungen im Oriente, während der Jahre 1843 und 1844* (Weimar: Druck und Verlag des Landes-Industrie-Comptoirs, 1828).

42 Brian Vickery, *Scientific Communication in History* (Lanham, MD: Scarecrow Press, 2000), 77.

43 L. Marsh, *Potted History of the Organisation(s) which Evolved to become the BSBI*, https://bsbi.org/wp-content/uploads/dlm_uploads/2022/03/BSBI-potted-history.pdf.

44 Francesc Martorell i Peña, *Apuntes arqueológicos de d. Francisco Martorell y Peña* (Barcelona: V. Dorca, 1979). He was not a trained botanist but a trader and collector of all things natural historical. For an account of his Moroccan travels, see A. González Bueno, *Les Campanyes Botàniques de Pius Font i Quer al Nord d'Àfrica* (Barcelona: Institut Botànic de Barcelona, 1988).

45 F. A. Stafleu, "A Century of Botanical Congresses," in *Proceedings of the XI International Botanical Congress*, University of Washington, Seattle, USA, August 24–September 2, 1969, ed. R. C. Starr (XI International Botanical Congress, Inc., 1970), 9–21.

46 Alphonse de Candolle, *Lois de la Nomenclature Botanique* (Paris: V. Masson et fils, 1867), 3.

47 de Candolle, *Lois*, 4.

48 de Candolle, *Lois*, 14.

49 de Candolle, *Lois*, 30–1.

50 de Candolle, *Lois*, 31.

51 This Code was published in French, English, and German in 1912 as *Règles internationales de la nomenclature botanique. International rules of botanical nomenclature. Internationale Regeln der botanischen Nomenclatur* (Jena: G. Fischer, 1912).

52 *Règles internationales de la nomenclature botanique*, 32.
53 *Règles internationales de la nomenclature botanique*, 40.
54 H. Trimen, "Botanical Nomenclature: Editor's response to A. de Candolle's Letter to the Journal on the Subject of Botanical Nomenclature," *Journal of Botany* 15 (1877): 243.
55 "The American Journal of Botany," *Science* 39 (1914): 570.
56 C. A. Weatherby, "Botanical Nomenclature Since 1867," *The American Journal of Botany* 36 (1949): 5–7.
57 International Botanical Congress of Cambridge, *International Rules of Botanical Nomenclature* (*IRBN*) (Jena: Verlag von Gustav Fischer, 1935), 11. The last Code in which this rule was enforced is the 2006 Vienna Code, which moved the rule to Article 36.1 and changed the wording so that it now states, "a name of a new taxon (algal and all fossil taxa excepted) must, in order to be validly published, be accompanied by a Latin description or diagnosis or by a reference to a previously and effectively published Latin description or diagnosis."
58 In the *ICN*, Linnaeus' binary epithets are known as *binary combinations*, but apart from this, they are almost identical in form to his original nomenclatural recommendations.
59 Stearn, *Botanical Latin*, 143.
60 Camp, et.al., *Proposed Changes in the International Rules of Botanical Nomenclature*. New York Botanical Garden and the American Society of Plant Taxonomists (Lancaster, PA: The Science Press, 1949).
61 W. H. Camp et al., "Proposed Changes in the International Rules of Botanical Nomenclature: A Report by the Central Committee on Nomenclature of the American Society of Plant Taxonomists," *Brittonia* 7 (1949): 35.
62 H. W. Rickett, "Some Changes in the International Rules of Botanical Nomenclature Made at Stockholm in 1950," *Bulletin of the Torrey Botanical Club* 78, no. 1 (1951): 6–7.
63 Rickett, "Some Changes," 8.
64 Rickett, "Some Changes," 9.
65 C. H. S. Kabuye, "Comments on Latin Descriptions," *Taxon* 39 (1990): 240.
66 Tarciso Filgueiras, "In Defense of Latin for Describing New Taxa," *Taxon* 46 (1997): 747.
67 A. J. G. H. Kostermans, "Plea to Kill a Holy Cow," *Taxon* 39 (1990): 234–5.
68 Kostermans, "Plea," 234.
69 Mohammad Nazeer Chaudhri, "The Sacred Cow and the Plight of Taxonomy in the Third World Countries," *Taxon* 40 (1991): 92.
70 Mohammad Nazeer Chaudhri, "(221) Proposal to Add English to the Latin Requirement for Validating Names," *Taxon* 41 (1992): 785.
71 Secretariat of the Convention on Biological Diversity, *Convention on Biological Diversity: Text and Annexes* (Montreal: United Nations Environment Programme, 2011), 1.

72 Secretariat, *Convention on Biological Diversity*, 7(j).

73 Gideon Smith and Estrela Figueiredo, "Responsible Species Description: A Change of Attitude is Needed to Facilitate and Improve Access to Biological Material," *Taxon* 60 (2011): 1549.

74 Secretariat, *Convention on Biological Diversity*, 1.

75 John McNeill, "Latin, the Renaissance Lingua Franca, and English, the 20th Century Language of Science: Their Role in Biotaxonomy," *Taxon* 46 (1997): 753–4. It is worth pointing out that McNeill also takes the time in this article to argue that botanical Latin is an evolving, living language, as evidenced by the particular style and syntax that developed over time.

76 Renato Goldenberg and André Amorim, "Physeterostemon (Melastomataceae): A New Genus and Two New Species from the Bahian Atlantic Forest, Brazil," *Taxon* 55 (2006): 966.

77 The authors make sure to cite Dr. William A. Rodrigues for the Latin.

78 Also known also as the Melbourne Code. The most recent code is the Madrid Code, 2025. The Code is also now conventionally referred to as the ICN (International Code of Nomenclature for algae, fungi, and plants).

79 Secretariat, *Convention on Biological Diversity*, 1.

5 Current Debates and Visual Communication

1 Carl Linnaeus, *Genera Plantarum* (Paris, M. A. David, 1743), iii.

2 https://www.iapt-taxon.org/nomen/pages/main/art_32.html.

3 There have been several instances of common or indigenous names being preserved and encouraged, usually outside of the scope of the IBC. The 1981 Wildlife and Countryside Act in the UK required the use of an English common name for all endangered species, while in Australia, the use of indigenous names is left up to the practitioner.

4 Gideon Smith and Estrela Figueiredo, "'*Rhodes*-' Must Fall: Some of the Consequences of Colonialism for Botany and Plant Nomenclature," *Taxon* 71 (2022): 1–5.

5 Henry Cranke Andrews, *The Botanist's Repository for New and Rare Plants* (London: The Author, 1799–1828), vol. 1 (1800), 291–2.

6 https://www.iapt-taxon.org/nomen/pages/main/preamble.html. The 2025 Code retains these statements.

7 Carl Linnaeus, *Philosophia Botanica* (Stockholm: Godofr. Kiesewetter, 1751), 208.

8 Sergei Mosyakin, "If '*Rhodes*-' Must Fall, Who Shall Fall Next?," *Taxon* 71 (2022a): 251.

9 Kevin Thiele et al., "Taxonomists have an Opportunity to Rid Botanical Nomenclature of Inappropriate Honorifics in a Structured and Defensible Way," *Taxon* 71 (2022): 1152.

10 The example provided involves a racial slur that is unfortunately preserved in several different genera and species names. The formal recommendation involves simply changing the spelling so that the slur is no longer a slur, but this author sees this as insufficient, as the removal of one letter from a litany of names bearing the term does not somehow erase its effects. For discussion of this particular term, see Gideon Smith and Estrela Figueiredo, "Proposal to Add a New Article 61.6 to Permanently and Retroactively Eliminate Epithets with the Root *caf[e]r-* or *caff[e]r-* from the Nomenclature of Algae, Fungi and Plants," *Taxon* 70 (2021): 1395–6.
11 Nicholas J. Turland, "From the *Shenzhen Code* to the *Madrid Code*: New Rules and Recommendations for Naming Algae, Fungi, and Plants," *American Journal of Botany* 112 (2025): 1–6.
12 The 2025 Madrid Code is published here: https://www.iaptglobal.org/_functions/code/madrid.
13 Sergei Mosyakin, "Proposals to Amend Recommendation 7A on Deposition of Type material in Institutions of Countries of Origin, and to Add a new Recommendation 51A Regarding Avoiding Potentially Inappropriate or Unacceptable Names of Taxa," *Taxon* 70 (2021): 1379.
14 Turland, "From the *Shenzhen Code* to the *Madrid Code*," 5.
15 Shane Wright and Len Gillman, "Replacing Current Nomenclature with Pre-Existing Indigenous Names in Algae, Fungi and Plants," *Taxon* 71 (2022): 6.
16 This is because at the species level, there is much less work involved in replacing a name; at the family level, for example, hundreds or even thousands of taxa would need their names altered.
17 Vera Hayova et al., "Prop. 221," *Taxon* 72 (2023): 455.
18 Clement Earp and Sergei Mosyakin, "(374) Proposal to Amend Article 38.3 to Clarify what Local, Indigenous or Traditional Information is to be Accepted in a Validating Description or Diagnosis," *Taxon* 72 (2023): 1167.
19 Nancy Turner, "Appendix 2B. Names of Selected Native Plant Species in Indigenous Languages of Northwestern North America," in *Ancient Pathways, Ancestral Knowledge: Ethnobotany and Ecological Wisdom of Indigenous Peoples of Northwestern North America* (Toronto: McGill-Queen's University Press, 2014), 491.
20 Karen Reeds, "Leonardo da Vinci and Botanical Illustration: Nature Prints, Drawings, and Woodcuts ca. 1500," in *Visualizing Medieval Medicine and Natural History, 1200–1550*, ed. J. A. Givens, Karen Reeds, and A. Touwaide (Aldershot: Ashgate, 2006), 208.
21 https://www.iapt-taxon.org/nomen/pages/main/art_40.html#Art40.4.
22 https://www.iapt-taxon.org/nomen/pages/main/art_40.html#Art40.5. Both of these have been slightly rephrased and moved to Article 40.6 in the Madrid Code.
23 Barbara Wittman, "Outlining Species: Drawing as a Research Technique in Contemporary Biology," *Science in Context, Supplement Knowledge in the Making* 26 (2013): 364.

24 Bruno Latour, "Drawing Things Together," in *Representation in Scientific Practice*, ed. Michael Lynch and Steve Woolgar (Cambridge, MA: MIT Press, 1990), 7.

25 Barbara Wittman, "Outlining Species," 364.

26 This discussion is not meant as a standard historical overview of botanical art and artists. There are already several excellent texts that discuss the processes and techniques of botanical illustration throughout time, e.g., Blunt and Stearn, *The Art of Botanical Illustration* (2021), Martyn Rix, *The Art of the Botanist* (1981), and Andrew Griebeler, *Botanical Icons* (2024). Historically, there were also other methods besides illustration for transmitting visual information about plants, such as wax models and the famous Ware collection of glass flowers at Harvard, made by the father-son duo Leopold and Rudolf Blaschka. This technique is still found in non-scientific, social-cultural, and artistic practices by such glass artists as Goldie Poblador, who creates glass sculptures of endangered plant species indigenous to the Philippines.

27 Fielding Garrison, "Herbals and Bestiaries," *Bulletin of the New York Academy of Medicine* 7 (1931): 893.

28 Generally speaking, Pliny's text itself was not illustrated. See, however, Lilian Armstrong ("The Illustration of Pliny's *Historia Naturalis*," 1983) for an analysis of three illuminated manuscripts of the *Historia Naturalis*, ranging from *c.* 1300 to the early fifteenth century (El Escorial, Biblioteca Real de San Lorenzo delEscorial, MS R.I.5; Turin, Biblioteca Nazionale Universitaria, MSS 1.1.24-1; Parma, Biblioteca Palatina, MS Parm. 1278 [H. H.1. 6]; and 6 Milan, Biblioteca Ambrosiana, MS E. 24). Most of the illustrations amount to initials, presentation scenes, landscapes, and miniatures. The dating of these earliest known illustrated Pliny manuscripts corresponds to the proliferation of illustrated herbals and medical texts in Italy, and there is little in these four manuscripts that is original or specific to Pliny.

29 Martyn Rix, *The Art of the Botanist* (Guildford: Lutterworth Press, 1981), 9 notes that even in Thomas Johnson's 1633 edition of John Gerard's *The Herball, or, General Historie of Plantes*, plates were citing Crateuas as the origin of the illustrations, as in the plant *Coronopus* on pg. 1190. The illustration itself (possibly a *Lotus* species) is copied from the Dioscorides ms, Codex Vindobonensis (aka the Codex Aniciae Julianae, now part of the collection of the Österreichische Nationalbibliothek in Vienna), which is one of the three believed to have copies of Crateuas' original illustrations.

30 John Lowden, "The Transmission of 'Visual Knowledge' in Byzantium through Illuminated Manuscripts: Approaches and Conjectures," in *Literacy, Education and Manuscript Transmission in Byzantium and Beyond*, ed. Catherine Holmes and Judith Waring (Leiden: Brill, 2002) applies this specifically to Byzantine religious art.

31 Andrew Griebeler, *Botanical Icons: Critical Practices of Illustration in the Premodern Mediterranean* (Chicago, IL: University of Chicago Press, 2024), 2.

32 See Vera S. Rutz, *Il Giardino Magico degli Alchimisti* (Milano: Il Polifilo, 2000) for the Italian alchemical herbal tradition.
33 Otto Pächt, "Early Italian Nature Studies and the Early Calendar Landscape," *Journal of the Warburg and Courtauld Institutes* 13, no. 1/2 (1950): 28.
34 Agnes Arber, *Herbals, their Origin and Evolution* (Cambridge: Cambridge University Press, 1912 [1986]), 319–38.
35 Griebeler, *Botanical Icons*, 2.
36 Lorraine Daston, "On Scientific Observation," *Isis* 99 (2008): 103.
37 Lorraine Daston, "Taking Note(s)," *Isis* 95 (2004): 444.
38 Griebeler's argument challenges this traditional approach, assimilating and summarizing (*Botanical Icons*, 2) Lowden's stance, i.e., that premodern herbal illustrations are worthy of study on their own, as a tradition separate from that of the texts they accompanied.
39 This is worth noting, since Weidtiz's career was defined primarily by his skill in woodblock art. The actual engraver of the woodblocks for Brunfels is unknown.
40 Walther Rytz, ed., *Pflanzenaquarelle des Hans Weiditz aus dem Jahre 1529; die Originale zu den Holzschnitten im Brunfels'schen Kräuterbuch* (Bern: Haupt, 1936). Not all of the watercolors were recovered, with only some fifteen being reproduced. One of the idiosyncrasies of these reproductions is that the editor has provided the Linnaean nomenclature for each herb, rather than their names in Brunfels.
41 See Hieronymus Bock's 1539 *Kreüterbuch*, fol. 152, in which he refers to *Gauchblům* as *Wild Kress*, which is precisely the nomenclature Leonhard Fuchs would suggest in 1542. This implies that the plant was well-known in Germany, where, despite the paucity of Brunfels' textual description, other authors were able to make a tentative identification based primarily on the illustration.
42 Rix, *Art of the Botanist*, 28.
43 Rix, *Art of the Botanist*, 30. For a reproduction of *Kuchenschell*, Rix has the following text: "*Right: Pasque flower (Anemone pulsatilla), woodcut from Herbarum Vivae Eicones (1530) by Otto Brunfels.*"
44 Lucie Čermáková and Jana Černá. "Naked in the Old and the New World: Differences and Analogies in Descriptions of European and American *herbae nudae* in the Sixteenth Century," *Journal of the History of Biology* 51 (2018): 71.
45 Norbert Flechsig, *Schottus adversus Egenolphum. Der erste "Urheberrechtsstreit" vor dem Reichskammergericht 1533/34. Nachdruckshutz gestern und heute* (Passau-Wien: MUR Verlag, 2017).
46 *vff fünff jar, nit nachzutrucken ac. bey der pen fünff marck lötigs golds.*
47 Sachiko Kusukawa, *Picturing the Book of Nature: Image, Text, and Argument in Sixteenth-Century Human Anatomy and Medical Botany* (Chicago, IL: University of Chicago Press, 2012), 107.

48 Sachiko Kusukawa, "Leonhart Fuchs on the Importance of Pictures," *Journal of the History of Ideas* 58 (1997): 404.

49 Arber, *Herbals*, 327.

50 Florike Egmond, *Eye for Detail: Images of Plants and Animals in Art and Science, 1500–1630* (London: Reaktion Books, 2017), 126.

51 Egmond, *Eye for Detail*, 128.

52 Maize and similar grains were mostly all called "granoturco."

53 Fuchs, *De Historia Stirpium*, 762ff. Morgan Library copy.

54 The Morgan copy has white flowers and the Oak Spring Garden copy has pink.

55 Fuchs, *De Historia Stirpium*, 391. The Morgan Library copy has pink, while the OSGF copy has blue. In comparison, the painted copy of Brunfels in the Peter Goop collection is far superior, both in the variety of pigments and the overall skill.

56 Peter Treveris, *The Grete Herball: Whiche Geueth Parfyt Knowlege and Under Standyng of all Maner of Herbes & There Gracyous Vertues whiche God hath Ordeyned for our Prosperous Welfare and Helth* (London: Peter Treveris, 1526), ch. 304.

57 Gessner died of plague in 1565, leaving the work unfinished. It was not published until 1754.

58 Daston, "Taking Note(s)," 444.

59 This simply means that they are unsigned. We do know, however, that an English botanist named Thomas Penny was staying with Gessner in 1565 and did some work on the herbal. For an account of Penny's travels in Switzerland and France, see Marie Addyman, "A Botanical Collector Abroad: Contextualizing Thomas Penny's Travels in Switzerland and France, 1565–1568," *Journal of the History of Collections* 33 (2021): 163–74.

60 Heinrich Zoller and Martin Steinmann, eds., *Conradi Gesneri Historia Plantarum* (Zurich: Urs Graf-Verlag, 1987/91), vol. 1, 17.

61 Egmond, *Eye for Detail*, 184.

62 There are a few other natural historical organisms, specifically fauna, depicted in the *Historia Plantarum*, such as fol. 305v, which is labeled as *Toreau sauvage*, but is actually *Bison bison* L., the American bison. Gessner never saw one in real life and it is unlikely he ever saw a preserved specimen, either. His illustration therefore is just slightly off, much like the beasts in medieval bestiaries.

63 Similarly, there is an illustration of corn (*Zea mays* L.), which shows an ear of corn partially husked, revealing the multiple colors of the corn cob.

64 Sachiko Kusukawa, "Image, Text and 'Observatio': The 'Codex Kentmanus'," *Early Science and Medicine* 14 (2009): 449–50.

65 The editors of the reproduction refer to these as *Aquarellierung teilweise* (partially painted in with watercolors").

66 This was also a technique used by Kentmann, who had assiduously recorded the actual sizes of plants, sometimes even noting that a plant or one of its parts was the

actual size shown. Kusukawa ("Image, Text," 453) concludes that his nature prints, e.g., the seedpod imprint of *Lunaria* on fol. 23r, were inspired by this insistence on recording dimensions.

67 Egmond, *Eye for Detail*, 11.

68 Egmond, *Eye for Detail*, 49.

69 One fruit is labeled *drei Früchte, eine aufgeschnitten mit mehreren Samen* (three fruits, one cut open with its seeds) and the other is labeled *zwei Früchte, eine aufgeschnitten mit Samen* (two fruits, one cut open with seeds).

70 Ghini's herbarium has not survived, but Cibo's is kept now in the R. Biblioteca di Roma.

71 Aldrovandi titled this *Paetum Indorum, Herba regina, Nicotiana seu Tabaco delle Indie*.

72 *Citrus auranthium* L., *Citrus limon* L., and *Citrus maxima* (Burm.).

73 A. Biancastella, *L'erbario di Ulisse Aldrovandi*. Titled *Maizum seu Triticum Bactrianum fructu albo et flavo* (Milano: F. Motta, 2003). Aldrovandi knew the difference between maize (American corn) and the grain called *Zea* in antiquity by Theophrastus.

74 Biancastella, *L'erbario*, 127. The dried specimen of this species bears the inscription *Mamey Indorum Persea quibusdam* (fol. 271).

75 See Chapter 3 for a discussion of Clusius' Latin.

76 Rembert Dodoens, *A Niewe Herball* (London: Gerard Dewes, 1578), 767.

77 Dodoens, *A Niewe Herball*, 764.

78 Fabio Colonna, *Phytobasanos, sive plantarum aliquot historia* (Naples: Apud Io. Iacobum Carlinum & Antonium Pacem, 1592), fol. 8v and pg. 92, respectively.

79 *HN* 27.99.

80 Prospero Alpini, *De Plantis Aegypti Liber Cur de Plantis in Aegypti Locis Provenientibus sit Agendum* (Venice: Apud Franciscum de Franciscis Senensem, 1592).

81 Nicolás Monardes, *Joyfull Newes out of the New-found Worlde* (London: Printed by E. Allde, 1596), fol. 75r.

82 He also notes that the new specimens seem to have been grown in moister areas (along the coasts) than the kind first found in Newe Spayne.

83 Jacques Contant and Paul Contant, *Les Oeuvres* (Poitiers: par J. Thoreau et la Vve d'A. Mesnier, 1628).

84 Antonio Donati, *Trattato de semplici* (Venice: Pietro Maria Bertano, 1631), fol. 252r.

85 Donati, *Trattato de semplici*, 21.

86 Charles Plumier, *Traité des Fougères de l'Amérique* (Paris: Royal Press, 1705). Plumier's influence on European knowledge of plants in the Americas cannot be overstated. He also produced *Description des plantes de l'amerique* in 1693, posthumously published as *Plantarum Americanarum*, 1755–60.

87 The species described was renamed a few times, but is now known as *Magnolia dodecapetala*.

88 Mark Catesby, *The Natural History of Carolina, Florida and the Bahama Island* (London: The Author, 1729–47), vol. 2, 60.

89 Niccolò Gavelli, *Storia distinta e curiosa del tabacco* (Pesaro: Nella Stamperia Gavelliana, 1758), 7.

90 Etienne-Francois Geoffroy, *Descriptions, Vertus, et Usages* (Paris: P. F. Didot le jeune, 1767), vol. 2, plate 231 (150).

91 Geoffroy, *Descriptions*, vol. 4, plate 514.

92 Geoffroy, *Descriptions*, vol. 1, plate 106 (6).

93 Wilfrid Blunt and William T. Stearn, *The Art of Botanical Illustration* (Woodbridge: ACC Art Books, 2021), 271.

94 Wittmann, "Outlining Species," 367.

95 Linnaeus, *Genera Plantarum*, vii.

Conclusion

1 Peter Burke, *What is the History of Knowledge?* (Cambridge: Polity Press, 2016).

2 In the context of this book, it is ironic that Bacon later translated this work into Latin under the title, *De Augmentis Scientiarum* (1623).

3 Auguste Comte, *The Positive Philosophy of Auguste Comte*, trans. Harriet Martineau (London: J. Chapman, 1853), vol. 1, 1.

4 Karl Mannheim, *Ideology and Utopia: An Introduction to the Sociology of Knowledge*, trans. Louis Wirth and Edward Shils (San Diego, CA: Harcourt Brace Jovanovich, 1936).

5 Burke, *What is the History of Knowledge?*, 4.

6 Burke, *What is the History of Knowledge?*, 5.

7 Thomas Kuhn, *The Structure of Scientific Revolutions* (Chicago, IL: University of Chicago Press, 1962 [2012]), 79–80.

8 Brian Vickery, *Scientific Communication in History* (Lanham, MD: Scarecrow Press, 2000), 123.

9 Michel Foucault, *The Order of Things: An Archaeology of the Human Sciences* (London: Tavistock, 1970), 50.

10 Foucault, *The Order of Things*, 326.

11 Foucault, *The Order of Things*, 50.

12 Jürgen Habermas, *Theory of Communication Action*, vol. 2, trans. T. McCarthy (Boston, MA: Beacon Press, 1987), 136.

13 See, e.g., Jasny et al. (2011) and National Academies of Sciences (2019) for thorough discussions of the issue. Franklin and Laymon (2021), however, argue that a valid

experiment does not actually need to be replicated. They point out that this so-called "gold standard" requirement in fact assumes that raw data are somehow free of theory. No matter the structure of an experiment, there is always an identifiable theoretical base in the results, and replication may have the unintended effect of confirming the biases and assumptions embedded in the original experiment. They therefore pose a series of strategies for evaluating the validity of an experiment without assuming the necessity of replication.

14 Lorraine Daston, "On Scientific Observation," *Isis* 99 (2008): 98.

15 Mikhail Bakhtin, *The Dialogic Imagination* (Austin, TX: University of Texas Press, 1993), 5.

16 Ibid.

17 Jordi Cat, *Fuzzy Pictures as Philosophical Problem and Scientific Practice: A Study of Visual Vagueness* (Cham: Springer, 2017), 4.

18 The most commonly used semantics (*STT*, the Semantic Theory of Truth) was developed by Alfred Tarski in the 1930s.

19 Not for the first time. He had published in 1930 a paper on a 3-valued modal logic, which he had first developed in 1920. See J. Łukasiewicz, "Philosophische Bemerkungen zu mehrwertigen Sytemen des Aussagenkalküls," *Comptes Rendus des Séances de la Sociétés des Sciences et des Lettres de Varsovie* xxiii (1930): 3.

20 J. Łukasiewicz, *Elements of Mathematical Logic* (Oxford: Pergamon Press, 1963), 67–8.

21 There are a couple of steps here, involving inserting the conjunction of A and not-A ($A \wedge \neg A$) into a conditional argument form, *modus ponens*, to obtain the conclusion that, from any instantiation of ($A \wedge \neg A$), one can obtain any conclusion.

22 Łukasiewicz, *Elements of Mathematical Logic*, 68.

23 Łukasiewicz, *Elements of Mathematical Logic*, 69–77 for this discussion and his methodology.

24 Łukasiewicz, *Elements of Mathematical Logic*, 78.

25 Bruno Latour, "Drawing Things Together," in *Representation in Scientific Practice*, ed. Michael Lynch and Steve Woolgar (Cambridge, MA: MIT Press, 1990), 2.

26 Latour, "Drawing Things Together," 3.

27 C. Hoffmann and B. Wittmann, "Introduction: Knowledge in the Making: Drawing and Writing as Research Techniques," *Science in Context* 26 (2013): 203.

28 Alan G. Gross and Joseph E. Harmon, *Science from Sight to Insight: How Scientists Illustrate Meaning* (Chicago, IL: University of Chicago Press, 2014), 1.

29 Gross and Harmon, *Science from Sight to Insight*, 1–2.

30 Gross and Harmon, *Science from Sight to Insight*, 81.

31 Wendy Makoon Geniusz, *Our Knowledge Is Not Primitive: Decolonizing Botanical Anishinaabe Teachings* (Syracuse, NY: Syracuse University Press, 2009), 4.

32 Geniusz, *Our Knowledge*, 6.

33 Frantz Fanon, *The Wretched of the Earth*, trans. Richard Philcox (New York: Grove Press, 1963 [2004]), 16.

34 Linda Tuhiwai Smith, *Decolonizing Methodologies: Research and Indigenous Peoples* (London: Zed Books, 2012), 75.

35 Compare this current debate with Tore Janson's statement (2004, 156) that Latin will likely continue to be used for botanical nomenclature because "there is simply nothing else that works as well." Janson does at least acknowledge that prior knowledge of Latin is helpful in understanding binomials and that many post-Linnaean names are inscrutable.

36 It is true that by replacing Latin names with English or indigenous names, the same issues might inevitably arise. In the case of the former, we are likely to encounter a new *lingua franca* eventually; in the latter case, it is also possible to argue that exclusivity is still a problem.

37 Carl Linnaeus, *Species Plantarum* (Stockholm: Laurentius Salvius, 1753), 2, 656.

Bibliography

Adams, J. N. "Introduction," in *Aspects of the Language of Latin Prose*, ed. T. Reinhardt, M. Lapidge, and J. N. Adams. Proceedings of the British Academy, 129 (Oxford: Oxford University Press, 2005), 1–36.

Addyman, Marie. "A Botanical Collector Abroad: Contextualizing Thomas Penny's Travels in Switzerland and France, 1565–1568," *Journal of the History of Collections* 33 (2021): 163–74.

Alpini, Prospero. *De Plantis Aegypti Liber Cur de Plantis in Aegypti Locis Provenientibus sit Agendum* (Venice: Apud Franciscum de Franciscis Senensem, 1592).

Alves, Ruy José Válka, Nílber Gonçalves da Silva, and Jorge Fontella Pereira. "Latin Shaken, not Stirred," *Taxon* 61 (2012): 246.

André, J. "Pline l'Ancien Botaniste," *Revue des Études Latines* 33 (1955): 297–318.

Andrews, Henry Cranke. *The Botanist's Repository for New and Rare Plants* (London: The Author, 1799–1828).

Arber, Agnes. *Herbals, Their Origin and Evolution: A Chapter in the History of Botany, 1470–1670* (Cambridge: Cambridge University Press, 1912 [1986]).

Armstrong, Lilian. "The Illustration of Pliny's *Historia Naturalis*: Manuscripts before 1430," *Journal of the Warburg and Courtauld Institutes* 46 (1983): 19–39.

Ashworth, William B. "Natural History and the Emblematic World View," in *Reappraisals of the Scientific Revolution*, ed. D. Lindberg and R. Westmann (Cambridge: Cambridge University Press, 1990), 303–32.

Ashworth, William B. "Emblematic Natural History of the Renaissance," in *Cultures of Natural History*, ed. N. Jardine, J. A. Secord, and E. C. Spary (Cambridge: Cambridge University Press, 1996), 17–37.

Bacon, Francis. *Of the Advancement of Learning* (1605 [2010 reprint, The Floating Press]).

Bakhtin, Mikhail. *The Dialogic Imagination* (Austin, TX: University of Texas Press, 1993).

Baranov, Andrei. *Basic Latin for Plant Taxonomists* (Lehre: J. Cramer, 1971).

Barbaro, Ermolao. *Hermolai Barbari Castigationes Plinianae et in Pomponium Melam / Edidit Giovanni Pozzi* (Passau: In aedibus Antenoreis, 1493).

Barbaro, Ermolao. *Historia Naturalis*, ed. Ermolao Barbaro, rev. Johannes Baptista Palmarius (Venice: Bernardinus Benalius, 1497).

Barbaro, Ermolao. *In Dioscoridem Corollariorum Libri Quinque* (Cologne: Ioan. Soterem, 1530).

Bartlett, H. "The Concept of the Genus: I. History of the Generic Concept in Botany," *Bulletin of the Torrey Botanical Club* 67 (1940): 349–62.

Bauhin, Kaspar. *Pinax Theatri Botanici* (Basil: Sumptibus & typis Ludovici Regis, 1623).

Bayton, Ross. *The Gardener's Botanical: An Encyclopedia of Latin Plant Names* (Princeton, NJ: Princeton University Press, 2020).

Bennett, Bradley C. "Doctrine of Signatures: An Explanation of Medicinal Plant Discovery or Dissemination of Knowledge?," *Economic Botany* 61 (2007): 246–55.

Bergemann, Lutz, Martin Dönike, Albert Schirrmeister, Georg Toepfer, Marco Walter, and Julia Weitbrecht. "Transformation: A Concept for the Study of Cultural Change," in *Beyond Reception: Renaissance Humanism and the Transformation of Classical Antiquity*, ed. and trans. Patrick Baker, Johannes Helmrath, and Craig Kallendorf. Transformationen der Antike, 62 (Berlin: De Gruyter, 2019), 9–26.

Betz, Hans D., and J. Scarborough. *The Greek Magical Papyri in Translation, Including the Demotic Spells* (Chicago, IL: University of Chicago Press, 1992).

Biancastella, A. *L'erbario di Ulisse Aldrovandi: natura arte e scienza in un tesorodel Rinascimento* (Milano: F. Motta, 2003).

Biggar, Henry P. *The Voyages of Jacques Cartier*. Publications of the Public Archives of Canada, 11 (Ottawa: F. A. Acland, 1924).

Bleichmar, Daniela. "Visual Culture in Eighteenth-Century Natural History" (PhD dissertation, Princeton University, 2004).

Blunt, Wilfrid, and William T. Stearn. *The Art of Botanical Illustration* (Woodbridge: ACC Art Books, 2021).

Boissier, Edmond. *Elenchus Plantarum Novarum Minusque Cognitarum quas in Itinere Hispanico* (Geneva: Typographia Lador et Ramboz, 1838).

Brand, John. "On the Latin Terms Used in Natural History," *Transactions of the Linnean Society* 3 (1979): 70–5.

Brasavola, Antonio Musa. *Examen Omnium Simplicium Medicamentorum* (Lyon: J. et F. Frellaeos, 1536).

Bridson, Gavin D. R., and James White. *Plant, Animal & Anatomical Illustration in Art & Science: A Bibliographical Guide from the 16th Century to the Present Day* (Winchester: St. Paul's Bibliographies in association with the Hunt Institute for Botanical Documentation, 1990).

Brockway, Lucille. *Science and Colonial Expansion: The Role of the British Royal Botanic Garden* (New Haven, CT: Yale University Press, 2002).

Brown, Richard D. *Knowledge Is Power: The Diffusion of Information in Early America, 1700–1865* (New York: Oxford University Press, 1989).

Brunfels, Otto. *Catalogus Illustrium Medicorum, sive de Primis Medicinae Scriptoribus, per Ot. Brunnfelsium* (Strassburg: Johann Schott, 1530).

Brunfels, Otto. *Herbarum Vivae Eicones* (Strassburg: Johann Schott, 1530–2).

Brunfels, Otto. *Epitome Medices Summam Totius Medicinae Complectens. Libellus vere Aureus Maxime Necessarius* (Antwerp: J. Montanus, 1540).

Burke, Peter. *A Social History of Knowledge: From Gutenberg to Diderot* (Cambridge: Polity Press, 2000).

Burke, Peter. *A Social History of Knowledge II: From the Encyclopédie to Wikipedia* (Cambridge: Polity Press, 2012).
Burke, Peter. *What is the History of Knowledge?* (Cambridge: Polity Press, 2016).
Camerarius, Joachim. *Hortus Medicus et Philosophicus* (Frankfurt: apud Iohannem Feyerabend, 1588).
Camp, W. H., H. W. Rickett, and C. A. Weatherby. "Proposed Changes in the International Rules of Botanical Nomenclature: A Report by the Central Committee on Nomenclature of the American Society of Plant Taxonomists," *Brittonia* 7 (1949): 1–51.
Candolle, Alphonse de. *Lois de la Nomenclature Botanique* (Paris: V. Masson et fils, 1867).
Candolle, Alphonse de. *La Phytographie, ou l'Art de Décrire les Végétaux* (Paris: G. Masson, 1880).
Carey, Sorcha. *Pliny's Catalogue of Culture: Art and Empire in the* Natural History (Oxford: Oxford University Press, 2003).
Carter, Jonathan. *Travels Through the Interior Parts of North America, In the Years 1766, 1767, and 1768* (London: C. Dilly, 1781).
Cat, Jordi. *Fuzzy Pictures as Philosophical Problem and Scientific Practice: A Study of Visual Vagueness* (Cham: Springer, 2017).
Catesby, Mark. *The Natural History of Carolina, Florida and the Bahama Islands* (London: The Author, 1729–47).
Čermáková, Lucie, and Jana Černá. "Naked in the Old and the New World: Differences and Analogies in Descriptions of European and American *herbae nudae* in the Sixteenth Century," *Journal of the History of Biology* 51 (2018): 69–106.
Cesalpino, Andrea. *De Plantis Libri xvi* (Florence: Georgius Marescottus, 1583).
Charlet, Jean-Louis, ed. *Deux Pièces de la Controverse Humaniste sur Pline: N. Perotti, Lettre à Guarnieri, C. Vitelli, Lettre à Partenio di Salò: Éditon Critique et Commentaire* (Sassoferrato: Istituto Internazionale di Studi Piceni, 2003).
Chaudhri, Mohammed Nazeer. "The Sacred Cow and the Plight of Taxonomy in the Third World Countries," *Taxon* 40 (1991): 92–3.
Chaudhri, Mohammed Nazeer. "(221) Proposal to Add English to the Latin Requirement for Validating Names," *Taxon* 41 (1992): 785.
Choate, Helen. "The Earliest Glossary of Botanical Terms; Fuchs 1542," *Torreya* 17 (1917): 186–201.
Clark, Albert C. "The Reappearance of the Texts of the Classics," *Transactions of the Bibliographical Society New Series* 2 (1921): 13–42.
Clusius, Carolus. *Caroli Clusii Atrebatis Rariorum Plantarum Historia, Quae Acceserint Proxima Pagina Docebit* (Antwerp: Ex Officina Plantiniana, Johann Moretus, 1601).
Cohn, Bernard. *Colonialism and Its Forms of Knowledge: The British in India* (Princeton, NJ: Princeton University Press, 1996).
Collenuccio, Pandulfo. *Pliniana Defensio Adversus Nicolai Leoniceni Accusationem* (Ferrera: Belfort, André, 1493).

Colonna, Fabio. *Phytobasanos, sive plantarum aliquot historia* (Naples: Apud Io. Iacobum Carlinum & Antonium Pacem, 1592).

Comte, Auguste. *The Positive Philosophy of Auguste Comte*, trans. Harriet Martineau (London: J. Chapman, 1853).

Contant, Jacques, and Paul Contant. *Les Oeuvres* (Poitiers: par J. Thoreau et la Vve d'A. Mesnier, 1628).

Convegno di Studi in Occasione del Quinto Centenario della Morte dell'Umanista Ermolao. Una famiglia Veneziana nella storia: I Barbaro: Atti del Convegno di Studi in occasione del Quinto Centenario della morte dell'umanista Ermolao, Venezia, 4–6 Novembre 1993/Raccolti da Michela Marangoni e Manlio Pastore Stocchi (Venezia: Istituto Veneto di Scienze, Lettere ed Arti, 1996).

Cooper, Alix. *Inventing the Indigenous: Local Knowledge and Natural History in Early Modern Europe* (Cambridge: Cambridge University Press, 2009).

Copeland, Rita. "Naming, Knowing, and the Object of Language in Alexander Neckam's Grammar Curriculum," *The Journal of Medieval Latin* 20 (2010): 38–57.

Daston, Lorraine. "Taking Note(s)," *Isis* 95 (2004): 443–8.

Daston, Lorraine. "On Scientific Observation," *Isis* 99 (2008): 97–100.

Daston, Lorraine, and Peter Galison. *Objectivity* (New York: Zone, 2007).

Davies, Martin. "Making Sense of Pliny in the Quattrocento," *Renaissance Studies* 9 (1995): 240–57.

De Candolle, Alphonse. *La Phytographie, ou l'Art de Décrire les Végétaux* (Paris: G. Masson, 1880).

De la Cruz, Martín. *Libellus de Medicinalibus Indorum Herbis* [Cruz-Badiano Codex], trans. Juan Badiano (México: Instituto Nacional de Antropología e Historia, 1552).

Dodoens, Rembert. *A Niewe Herball* (London: Gerard Dewes, 1578).

Donati, Antonio. *Trattato de Semplici* (Venice: Pietro Maria Bertano, 1631).

Doody, Aude. *Pliny's Encyclopedia: The Reception of the Natural History* (Cambridge: Cambridge University Press, 2010).

Du Pinet, Antoine. *Historia Plantarum* (Lyon: apud Gabrielem Coterium, 1561).

Earp, Clement, and Sergei Mosyakin. "(374) Proposal to Amend Article 38.3 to Clarify what Local, Indigenous or Traditional Information is to be Accepted in a Validating Description or Diagnosis," *Taxon* 72 (2023): 1167–8.

Egmond, Florike. *The World of Carolus Clusius: Natural History in the Making, 1550–1610* (London: Pickering & Chatto, 2010).

Egmond, Florike. "Observing Nature: The Correspondence Network of Carolus Clusius (1526–1609)," in *Communicating Observations in Early Modern letters (1500–1675): Epistolography and Epistemology in the Age of the Scientific Revolution*, ed. Dirk van Miert (London: Warburg Institute, 2013).

Egmond, Florike. *Eye for Detail: Images of Plants and Animals in Art and Science, 1500–1630* (London: Reaktion Books, 2017).

Eisenstein, Elizabeth. *The Printing Press as an Agent of Change: Communications and Cultural Transformations in Early Modern Europe*, 2 vols. (Cambridge: Cambridge University Press, 1979).

Epstein, Marcelo, Ruth Spivak, and Donald Sprague, eds. *The Latin of Science* (Mundelein, IL: Bolchazy-Carducci Publishers, 2019).

Erichsen-Brown, Charlotte. *Medicinal and Other Uses of North American Plants: A Historical Survey with Special Reference to the Eastern Indian Tribes* (New York: Dover Publications, 1979).

Fabroni, Angelo. *Vitae Italorum Doctrina Excellentium qui Saeculis XVII et XVIII. Floruerunt* (Pisa: Carolus Ginesius, 1778).

Fanon, Frantz. *The Wretched of the Earth*, trans. Richard Philcox; with commentary by Jean-Paul Sartre and Homi K. Bhabha (New York: Grove Press, 1963 [2004]).

Fernald, Merritt L. *Gray's Manual of Botany: A Handbook of the Flowering Plants and Ferns of the Central and Northeastern United States and Adjacent Canada* (Portland, OR: Dioscorides Press, 1950).

Filgueiras, Tarciso. "In Defense of Latin for Describing New Taxa," *Taxon* 46 (1997): 747–9.

Findlen, Paula. *Possessing Nature: Museums, Collecting, and Scientific Culture in Early Modern Italy* (Berkeley, CA: University of California Press, 1994).

Flechsig, Norbert. *Schottus adversus Egenolphum. Der erste "Urheberrechtsstreit" vor dem Reichskammergericht 1533/34. Nachdruckshutz gestern und heute* (Passau-Wien: MUR Verlag, 2017).

Fleck, Ludwick. *Genesis and Development of a Scientific Fact* (Chicago, IL: University of Chicago Press, 1979).

Foucault, Michel. *Les Mots et les Choses: Une Archéologie des Sciences Humaines* (Paris: Gallimard, 1966).

Foucault, Michel. *The Order of Things: An Archaeology of the Human Sciences* (London: Tavistock, 1970).

Franklin, Allan, and Ronald Laymon. *Once Can Be Enough: Decisive Experiments, No Replication Required* (Cham: Springer, 2021).

French, R. "Pliny and Renaissance Medicine," in *Science in the Early Roman Empire*, ed. R. French and F. Greenaway (London: Croom Helm, 1986), 252–81.

Fuchs, Leonhart. *Errata Recentiorum Medicorum, LX Numero, Adiectis Eorundem Confutationibus* (Secerius, 1530).

Fuchs, Leonhart. *De Historia Stirpium Commentarii Insignes* (Basil: In officina Isingriniana, 1542).

Garrison, Fielding H. "Herbals and Bestiaries," *Bulletin of the New York Academy of Medicine* 7 (1931): 891–904.

Gavelli. Niccolò. *Storia distinta e curiosa del tabacco* (Pesaro: Nella Stamperia Gavelliana, 1758).

Geniusz, Wendy Makoons. *Our Knowledge Is Not Primitive: Decolonizing Botanical Anishinaabe Teachings* (Syracuse, NY: Syracuse University Press, 2009).

Geoffroy, Étienne-François. *Descriptions, Vertus, et Usages* (Paris: P. F. Didot le jeune, 1767).

Gerard, John. *The Herball, or, Generall historie of plantes* (London: John Norton, 1597).

Gibson, Roy, and Ruth Morello. *Pliny the Elder: Themes and Contexts*. Mnemosyne Supplements. Monographs on Greek and Latin Language and Literature (Leiden: Brill, 2011).

Goldenberg, Renato, and André M. Amorim. "*Physeterostemon* (Melastomataceae): A New Genus and Two New Species from the Bahian Atlantic Forest, Brazil," *Taxon* 55 (2006): 965–72.

González Bueno, A. *Les Campanyes Botàniques de Pius Font i Quer al Nord d'Àfrica* (Barcelona: Institut Botànic de Barcelona, 1988).

Goodyear, F. R. D. "Technical Writing," in *The Cambridge History of Classical Literature, vol. 2: Latin Literature*, ed. E. J. Kenney and W. V. Clausen (Cambridge: Cambridge University Press, 1982), 667–73.

Grafton, Anthony. *Magic and Technology in Early Modern Europe* (Washington, DC: Smithsonian Institution Libraries, 2005).

Greene, Edward L. *Landmarks of Botanical History* (Washington, DC: The Smithsonian Institution, 1909).

Griebeler, Andrew. *Botanical Icons: Critical Practices of Illustration in the Premodern Mediterranean* (Chicago, IL: University of Chicago Press, 2024).

Grimal, Pierre. *Les jardins romains à la fin de la République et aux deux premiers siècles de l'Empire; essai sur le naturalisme romain* (Paris: E. de Boccard, 1943).

Gross, Alan G., and Joseph E. Harmon. *Science from Sight to Insight: How Scientists Illustrate Meaning* (Chicago, IL: University of Chicago Press, 2014).

Guida, Augusto. "Plinio il Vecchio, un Postillato Polizianeo e un Progetto per Winkelmann," *Prometheus* 44 (2018): 280–93.

Habermas, Jürgen. *Theory of Communication Action*, vol. 2, trans. T. McCarthy (Boston, MA: Beacon Press, 1987).

Hall, Edwin. *Sweynheym & Pannartz and the Origins of Printing in Italy: German Technology and Italian Humanism in Renaissance Rome* (McMinnville, OR: Phillip J. Pirages, 1991).

Hardy, Gavin, and Laurence Totelin, *Ancient Botany* (New York: Routledge, 2016).

Harrison, Lorraine. *Latin for Gardeners* (Chicago, IL: University of Chicago Press, 2012).

Hayova, Vera, Ganno Boiko, and Sergei Mosyakin. "Prop. 221," *Taxon* 72 (2023): 455.

Heiberg, J. L., ed. *Paulus Aegineta, Corpus Medicorum Graecorum, IX* (Leipzig: B. G. Teubner, 1924), vol. II, 401–8.

Hellinga, Lotte. *Texts in Transit: Manuscript to Proof and Print in the Fifteenth Century* (Leiden: Brill, 2014).

Hemsley, William Botting. "The History of the Botanical Magazine," in *A New and Complete Index to the Botanical Magazine* (London: Lovell Reeve, 1906), v–lxiii.

Housman, Alfred E. "The Editing of Manilius," in *Collected Poems and Selected Prose*, ed. C. Ricks (London: Allen Lane, 1988), 372–94.

Hoffmann, C., and B. Wittmann. "Introduction: Knowledge in the Making: Drawing and Writing as Research Techniques," *Science in Context* 26 (2013): 203–13.

International Association for Plant Taxonomy. *Seventeenth International Botanical Congress.* "International Code of Botanical Nomenclature (Vienna Code). Adopted by the Seventeenth International Botanical Congress, Vienna, Austria, July 2005" (Vaduz: Gantner Verlag, 2006).

International Association for Plant Taxonomy. *Eighteenth International Botanical Congress.* "International Code of Botanical Nomenclature (Melbourne Code). Adopted by the Eighteenth International Botanical Congress, Melbourne, Australia, July 2011" (Vaduz: Gantner Verlag, 2012).

International Association for Plant Taxonomy. *Nineteenth International Botanical Congress.* "International Code of Nomenclature for Algae, Fungi, and Plants (Shenzhen Code). Adopted by the Nineteenth International Botanical Congress, Shenzhen, China, July 2017" (Vaduz: Gantner Verlag, 2018).

International Association for Plant Taxonomy. *Twentieth International Botanical Congress.* "International Code of Nomenclature for Algae, Fungi, and Plants (Madrid Code). Adopted by the Twentieth International Botanical Congress, Madrid, Spain, July 2024" (Chicago, IL: University of Chicago Press, 2025).

International Botanical Congress. *Règles internationales de la nomenclature botanique. International rules of botanical nomenclature. Internationale Regeln der botanischen Nomenclatur* (Jena: G. Fischer, 1912).

International Botanical Congress of Cambridge. *International Rules of Botanical Nomenclature, adopted by the Botanical Congresses of Vienna, 1905, and Brussels, 1910, Revised by the International Botanical Congress of Cambridge, 1930* (Jena: Verlag von Gustav Fischer, 1935).

Irmscher, Christoph. *The Poetics of Natural History* (New Brunswick, NJ: Rutgers University Press, 2019).

Jackson, Ralph. *Doctors and Diseases in the Roman Empire* (London: British Museum Publications, 1988).

Janson, Tore. *A Natural History of Latin* (Oxford: Oxford University Press, 2004).

Jasny, Barabara R., Gilbert Chin, Lisa Chong, and Dacha Vignieri. "Again, and Again, and Again . . .," *Science* 334 (2011): 1225.

Jensen, Kristian. "The Humanist Reform of Latin and Latin Teaching," in *The Cambridge Companion to Renaissance Humanism*, ed. J. Kraye. Cambridge Companions to Literature and Classics (Cambridge: Cambridge University Press, 1996), 63–81.

Jones, Evan T. "The Matthew of Bristol and the Financiers of John Cabot's 1497 voyage to North America," *English Historical Review* 121 (2006): 778–95.

Jones, Peter. *Quid pro Quo: What the Romans Really Gave the English Language* (London: Atlantic Books, 2016).

Kabuye, C. H. S. "Comments on Latin Descriptions," *Taxon* 39 (1990): 240.

Kallendorf, Craig. *Printing Virgil: The Transformation of the Classics in the Renaissance* (Leiden: Brill, 2020).

Kemp, Martin. *The Science of Art* (New Haven, CT: Yale University Press, 1990).

King, Margaret. *Venetian Humanism in an Age of Patrician Dominance* (Princeton, NJ: Princeton University Press, 1986).

King, V. H. "An Investigation of Some Astronomical Excerpts from Pliny's Natural History Found in Manuscripts of the Earlier Middle Ages" (B. Litt. thesis, Oxford, 1969).

Knight, Leah. *Of Books of Botany in Early Modern England* (London: Routledge, 2016).

Koch, Karl. *Wanderungen im Oriente, während der Jahre 1843 und 1844* (Weimar: Druck und Verlag des Landes-Industrie-Comptoirs, 1828).

Kostermans, A. J. G. H. "Plea to Kill a Holy Cow," *Taxon* 39 (1990): 234–5.

Kristeller, P. O. "Humanism and Scholasticism Renaissance," *Byzantion* 17 (1944–5): 346–74.

Kühn, C. G., ed. *Claudii Galeni Opera Omnia* (Leipzig: C. Cnoblochii, 1830), 19: 721–47.

Kuhn, Thomas. *The Structure of Scientific Revolutions* (Chicago, IL: University of Chicago Press, 1962 [2012]).

Kusukawa, Sachiko. "Leonhart Fuchs on the Importance of Pictures," *Journal of the History of Ideas* 58 (1997): 403–27.

Kusukawa, Sachiko. "Image, Text and 'Observatio': The 'Codex Kentmanus'," *Early Science and Medicine* 14 (2009): 445–75.

Kusukawa, Sachiko. *Picturing the Book of Nature: Image, Text, and Argument in Sixteenth-Century Human Anatomy and Medical Botany* (Chicago, IL: University of Chicago Press, 2012).

Kusukawa, Sachiko. "Drawing as an Instrument of Knowledge: The Case of Conrad Gessner," in *Vision and Its Instruments: Art, Science and Technology in Early Modern Europe*, ed. A. Payne (University Park, PA: Pennsylvania State University Press, 2015), 36–48.

Laird, Andrew. *The Epic of America: An Introduction to Rafael Landívar and the Rusticatio Mexicana* (London: Bloomsbury Academic, 2006).

Latour, Bruno. "Visualization and Cognition: Drawing Things Together," in *Knowledge and Society: Studies in the Sociology of Culture, Past and Present*, vol. 6, ed. H. Kuklick and Elizabeth Long (Greenwich, CT: JAI Press, 1986), 1–40.

Latour, Bruno. "Drawing Things Together," in *Representation in Scientific Practice*, ed. Michael Lynch and Steve Woolgar (Cambridge, MA: MIT Press, 1990), 19–68.

Latour, Bruno. *Pandora's Hope: Essays on the Reality of Science Studies* (Cambridge, MA: Harvard University Press, 1999).

Latour, Bruno, and Steve Woolgar. *Laboratory Life: The Social Construction of Scientific Facts*. Sage Library of Social Research, 80 (Beverly Hills, CA: Sage, 1979).

León, Nicolás. *Catálogo Bibliográfico, Biográfico y Crítico de Autores y Escritos Referentes a Vegetales de México y Sus Aplicaciones, desde la Conquista hasta el Presente* (Mexico: Oficina Tip. de la Secretaría de Fomento, 1895).

Leonhardt, Jürgen. *Latin: The Story of a World Language* (Cambridge, MA: The Belknap Press of Harvard University Press, 2013).

Leoniceno, Niccolò. *De Plinii et Aliorum in Medicina Erroribus* (Ferrara: Rossi, L. de Grassis, A., 1492).

Leu, Urs B. "The Rediscovered Third Volume of Conrad Gessner's 'Historia Plantarum'," in *For the Sake of Learning: Essays in Honor of Anthony Grafton*, vol. 2, ed. A. Blair and A. Goeing (Leiden: Brill, 2016), 415–22.

Lindley, John. *The Elements of Botany, Structural, Physiological, & Medical* (London: Printed for Taylor and Walton, 1841).

Linnaeus, Carl. *Fundamenta Botanica quae Majorum Operum Prodromi Instar Theoriam Scientiae Botanices per Breves Aphorismos Tradunt* (Amsterdam: Apud Salomonem Schouten, 1736).

Linnaeus, Carl. *Critica Botanica in qua Nomina Plantarum Generica* (Leiden: Apud C. Wishoff, 1737).

Linnaeus, Carl. *Genera Plantarum: Eorumque Characteres Naturales, secundum Numerum, Figuram, Situm & Proportionem Omnium Fructificationis Partium* (Paris: M. A. David, 1743).

Linnaeus, Carl. *Philosophia Botanica: in qua Explicantur Fundamenta Botanica cum Definitionibus Partium, Exemplis Terminorum, Observationibus Rariorum, Adjectis Figuris Aeneis* (Stockholm: Godofr. Kiesewetter, 1751).

Linnaeus, Carl. *Species Plantarum, Exhibentes Plantas Rite Cognitas, ad Genera Relatas, cum Differentiis Specificis, Nominibus Trivialibus, Synonymis Selectis, Locis Natalibus, secundum Systema Sexuale Digestas* (Stockholm: Laurentius Salvius, 1753).

Lowden, John. "The Transmission of 'Visual Knowledge' in Byzantium through Illuminated Manuscripts: Approaches and Conjectures," in *Literacy, Education and Manuscript Transmission in Byzantium and Beyond*, ed. Catherine Holmes and Judith Waring (Leiden: Brill, 2002), 59–80.

Łukasiewicz, Jan. "Philosophische Bemerkungen zu mehrwertigen Sytemen des Aussagenkalküls," *Comptes Rendus des Séances de la Sociétés des Sciences et des Lettres de Varsovie* xxiii (1930): 3.

Łukasiewicz, Jan. *Elements of Mathematical Logic* (Oxford: Pergamon Press, 1963).

Mannheim, Karl. *Ideology and Utopia: An Introduction to the Sociology of Knowledge*, trans. Louis Wirth and Edward Shils (San Diego, CA: Harcourt Brace Jovanovich, 1936).

Marhold, K. "Typification of the Linnaean Names of the Genus *Cardamine* (Cruciferae)," *Botanical Journal of the Linnean Society* 121 (1996): 124–5.

Martorell i Peña, Francesc. *Apuntes Arqueológicos de d. Francisco Martorell y Peña, Ordenados por Salvador Sanpere y Miquel, Pub. por don Juan Martorell y Peña* (Barcelona: V. Dorca, 1879).

McHam, Sarah B. *Pliny and the Artistic Culture of the Italian Renaissance: The Legacy of the* Natural History (New Haven, CT: Yale University Press, 2013).

McNeill, John. "Latin, the Renaissance Lingua Franca, and English, the 20th Century Language of Science: Their Role in Biotaxonomy," *Taxon* 46 (1997): 751–7.

Mercati, Giovanni. *Per la cronologia della vita e degli scritti di Niccolò Perotti Arcivescovo di Siponto*. Studi e Testi, 44 (Roma: Biblioteca Apostolica Vaticana, 1925).

Merian, Maria Sibylla. *Metamorphosis Insectorum Surinamensium. Ofte verandering der Surinaamsche insecten* (Amsterdam: Voor den auteur, als ook by G. Valck, 1705).

Merula, Giorgio. *In Librum de Homine Martii Galeotti opus. Add: Epistolae; In Sapphus epistolam interpretatio; Emendationes Plinii; Emendationes Vergilii* (Venice: Johann von Köln and Manthen, Johann, 1470).

Miller, Perry, and Thomas H. Johnson, eds. *The Puritans: A Sourcebook of their Writings*, 2 vols. (New York: Harper & Row, 1963).

Mitten, William. *Musci Austro-Americani, Enumeratio Muscorum Omnium Austro-Americanorum Auctori huiusque Cognitorum* (London: Longmans, Green, Reader, & Dyer, 1869 [Monographs in Systematic Botany from the Missouri Botanical Garden, 7; 1982 reprint]).

Monardes, Nicolás. *Joyfull Newes out of the New-found Worlde* (London: Printed by E. Allde, 1596).

Monfasani, John. "The First Call for Press Censorship: Niccolò Perotti, Giovanni Andrea Bussi, Antonio Moreto, and the Editing of Pliny's *Natural History*," *Renaissance Quarterly* 41 (1988): 1–31.

Monfasani, John. *Language and Learning in Renaissance Italy: Selected Articles* (Aldershot: Variorum, 1994).

Morton, A. G. *History of Botanical Science: An Account of the Development of Botany from Ancient Times to the Present Day* (London: Academic Press, 1981).

Morton, A. G. "Pliny on Plants," in *Science in the Early Roman Empire: Pliny the Elder, His Sources and Influence*, ed. R. French and F. Greenaway (London: Croom Helm, 1986), 86–97.

Mosyakin, Sergei. "Proposals to Amend Recommendation 7A on Deposition of Type material in Institutions of Countries of Origin, and to Add a new Recommendation 51A Regarding Avoiding Potentially Inappropriate or Unacceptable Names of Taxa," *Taxon* 70 (2021): 1379–80.

Mosyakin, Sergei. "If '*Rhodes-*' Must Fall, Who Shall Fall Next?," *Taxon* 71 (2022): 249–55.

Mosyakin, Sergei. "Defending Art. 51 of the *Code*: Comments on Smith & al.," *Taxon* 71 (2022): 1141–50.

Mowat, J. L. G., ed. *Alphita: A Medico-Botanical Glossary from the Bodleian Manuscript, Selden B.35* (Oxford, Clarendon Press, 1887).

National Academies of Sciences, Engineering, and Medicine. *Reproducibility and Replicability in Science* (Washington, DC: The National Academies Press, 2019).

Nauert, Charles. "Humanists, Scientists, and Pliny: Changing Approaches to a Classical Author," *The American Historical Review* 84 (1979): 72–85.

Nauert, Charles. "C. Plinius Secundus (*Naturalis Historia*)," *Catalogus Translationum et Commentariorum* IV (1980): 297–422.

Nickelsen, Kärin. *Draughtsmen, Botanists and Nature: The Construction of Eighteenth-Century Botanical Illustrations* (Dordrecht: Springer, 2006).

Norden, Edouard. *Die antike Kunstprosa vom VI. Jahrhundert v. Chr. bis in die Zeit der Renaissance* (Leipzig: Teubner, 1898).

Nutton, Vivian. "The Rise of Medical Humanism: Ferrara, 1464–1555," *Renaissance Studies* 11 (1997): 2.

Nutton, Vivian. *Ancient Medicine* (London: Routledge, 2012).

Nybakken, Oscar E. *Greek and Latin in Scientific Terminology* (Ames, IA: Iowa State College Press, 1959).

Ogilvie, Brian. *The Science of Describing: Natural History in Renaissance Europe* (Chicago, IL: University of Chicago Press, 2006).

Pächt, Otto. "Early Italian Nature Studies and the Early Calendar Landscape," *Journal of the Warburg and Courtauld Institutes* 13, no. 1/2 (1950): 13–47.

Paparazzo, Ernesto. "Philosophy and Science in the Elder Pliny's *Naturalis Historia*," in *Pliny the Elder: Themes and Contexts*. Mnemosyne Supplements. Monographs on Greek and Latin Language and Literature (Leiden: Brill, 2011), 89–112.

Pinkster, H. "The Language of Pliny the Elder," in *Aspects of the Language of Latin Prose*, ed. T. Reinhardt, M. Lapidge, and J. N. Adams. Proceedings of the British Academy, 129 (Oxford: Oxford University Press, 2005), 238–56.

Plumier, Charles. *Traité des Fougères de l'Amérique* (Paris: Royal Press, 1705).

Politian, Angelus. *Letters*, ed. and trans. Shane Butler (Cambridge, MA: Harvard University Press, 2006).

Pona, Giovanni. *Plantae seu Simpliciae* (Basil: Lazarus Zetzner, 1608).

Pozza, Neri. *La stampa degli incunaboli nel Veneto* (Vicenza: N. Pozza, 1984).

Preisendanz, K., ed. *Papyri Graecae Magicae; die griechischen Zauberpapyri* (Stuttgart: B. G. Teubner, 1928).

Ragep, F. J., S. P. Ragep, and S. J. Livesey. *Tradition, Transmission, Transformation: Proceedings of Two Conferences on Pre-Modern Science Held at the University of Oklahoma* (Leiden: Brill, 1996).

Ramirez Cabañas, J., ed. *Bernal Díaz del Castillo, Historia Verdadera de la Conquista de la Nueva España* (México: Pedro Robredo, 1944).

Reeds, Karen. "Leonardo da Vinci and Botanical Illustration: Nature Prints, Drawings, and Woodcuts ca. 1500," in *Visualizing Medieval Medicine and Natural History, 1200–1550*, ed. J. A. Givens, Karen Reeds, and A. Touwaide (Aldershot: Ashgate, 2006), 205–37.

Reeve, M. "The Editing of Pliny's Natural History," *Revue d'Histoire des Textes* 2 (2007): 107–79.

Reynolds, Leighton Durham. "The Elder Pliny," in *Texts and Transmission: A Survey of the Latin Classics*, ed. Leighton Durham Reynolds (Oxford: Clarendon Press, 1983), 307–16.

Reynolds, Leighton Durham, and N. G. Wilson. *Scribes and Scholars: A Guide to the Transmission of Greek and Latin Literature* (Oxford: Clarendon Press, 1974).

Rickett, H. W. "Some Changes in the International Rules of Botanical Nomenclature Made at Stockholm in 1950," *Bulletin of the Torrey Botanical Club* 78, no. 1 (1951): 1–10.

Rivinus, Augustus Quirinus. *Introductio Generalis in Rem Herbarium* (Lipsae: Christoph. Güntheri, 1690).

Rix, Martyn. *The Art of the Botanist* (Guildford: Lutterworth Press, 1981).
Rizzo, Silvia. *Il lessico filologico degli umanisti* (Roma: Edizioni di Storia e Letteratura, 1973).
Rutz, Vera S. *Il giardino magico degli alchimisti: Un erbario illustrato Trecentesco della Biblioteca Universitaria di Pavia e la sua tradizione* (Milano: Il Polifilo, 2000).
Rytz, Walther, ed. *Pflanzenaquarelle des Hans Weiditz aus dem Jahre 1529; die Originale zu den Holzschnitten im Brunfels'schen Kräuterbuch* (Bern: Haupt, 1936).
Schnalke, Thomas. "*Plantae Selectae*: The Flowering of Botanical Illustration in the 18th Century," in *The Pharmacy: Windows on History*, ed. Regine Pätzsch and Alain Touwaide (Basel: Editiones Roche, 1996), 195–209.
Scholderer, Victor. *Printers and Readers in Italy in the Fifteenth Century* (London, G. Cumberlege, 1949).
Secretariat of the Convention on Biological Diversity. *Convention on Biological Diversity: Text and Annexes* (Montreal: United Nations Environment Programme, 2011).
Short, Emma, and Alex George. *A Primer of Botanical Latin with Vocabulary* (Cambridge: Cambridge University Press, 2013).
Singer, Charles. "The Herbal in Antiquity and Its Transmission to Later Ages," *Journal of Hellenic Studies* 47 (1927): 1–52.
Smith, Gideon F., and Estrela Figueiredo. "Responsible Species Description: A Change of Attitude Is Needed to Facilitate and Improve Access to Biological Material," *Taxon* 60 (2011): 1549–51.
Smith, Gideon F., and Estrela Figueiredo. "Proposal to Add a New Article 61.6 to Permanently and Retroactively Eliminate Epithets with the Root *caf[e]r-* or *caff[e]r-* from the Nomenclature of Algae, Fungi and Plants," *Taxon* 70 (2021): 1395–6.
Smith, Gideon F., and Estrela Figueiredo. "'*Rhodes-*' Must Fall: Some of the Consequences of Colonialism for Botany and Plant Nomenclature," *Taxon* 71 (2022): 1–5.
Smith, Linda Tuhiwai. *Decolonizing Methodologies: Research and Indigenous Peoples* (London: Zed Books, 2012).
Smith, Pamela. *The Business of Alchemy: Science and Culture in the Holy Roman Empire* (Princeton, NJ: Princeton University Press, 1994).
Sprague, T. A. "The Herbal of Otto Brunfels," *Journal of the Linnean Society of London, Botany* 48 (1928): 79–124.
Stafleu, F. A. "A Century of Botanical Congresses," in *Proceedings of the XI International Botanical Congress*, University of Washington, Seattle, USA, August 24–September 2, 1969, ed. R. C. Starr (XI International Botanical Congress, Inc., 1970), 9–21.
Stearn, William S. *Botanical Latin: History, Grammar, Syntax, Terminology and Vocabulary* (London: Nelson, 1966).
Stearn, William S. *Botanical Latin: History, Grammar, Syntax, Terminology and Vocabulary*, 4th edition (Portland, OR: Timber Press, 1995).
Stearn, William S. *Dictionary of Plant Names for Gardeners* (Portland, OR: Timber Press, 2002).

"The American Journal of Botany," *Science* 39 (1914): 570.

Thiele, Kevin R., Gideon F. Smith, Estrela Figueiredo, and Timothy A. Hammer. "Taxonomists have an Opportunity to Rid Botanical Nomenclature of Inappropriate Honorifics in a Structured and Defensible Way," *Taxon* 71 (2022): 1151–4.

Treveris, Peter. *The Grete Herball: Whiche Geueth Parfyt Knowlege and Under Standyng of all Maner of Herbes & There Gracyous Vertues whiche God hath Ordeyned for our Prosperous Welfare and Helth* (London: Peter Treveris, 1526).

Trimen, H. "Botanical Nomenclature: Editor's Response to A. de Candolle's Letter to the Journal on the Subject of Botanical Nomenclature," *Journal of Botany* 15 (1877): 242–3.

Turland, Nicholas J. "From the *Shenzhen Code* to the *Madrid Code*: New Rules and Recommendations for Naming Algae, Fungi, and Plants," *American Journal of Botany* 112 (2025): 1–6.

Turner, Nancy J. "Appendix 2B. Names of Selected Native Plant Species in Indigenous Languages of Northwestern North America," in *Ancient Pathways, Ancestral Knowledge: Ethnobotany and Ecological Wisdom of Indigenous Peoples of Northwestern North America* (Toronto: McGill-Queen's University Press, 2014), 473–506.

Umachandran, Mathur, and Marchella Ward, eds. *Critical Ancient World Studies: The Case for Forgetting Classics* (London: Routledge, 2024).

Varey, Simon. *Searching for the Secrets of Nature: The Life and Works of Dr. Francisco Hernández* (Stanford, CA: Stanford University Press, 2000).

Vázquez, Francisco M., María Gutiérrez, and Soledad Ramos. "Diversity of the Gall Oak (*Quercus faginea* Lam.) in the Iberian Peninsula," *International Oaks: The Journal of the International Oak Society* 21 (2010): 121–36.

Vickery, Brian C. *Scientific Communication in History* (Lanham, MD: Scarecrow Press, 2000).

Wallace-Hadrill, Andrew. "Pliny the Elder and Man's Unnatural History," *GAR* 37 (1990): 80–96.

Weatherby, C. A. "Botanical Nomenclature Since 1867," *The American Journal of Botany* 36 (1949): 5–7.

Webster, Colin. "Heuristic Medicine: The Methodists and Metalepsis," *Isis* 106 (2015): 657–68.

Wellcome, Henry S. *Souvenir Cinchona Tercentenary Celebration and Exhibition at the Wellcome Historical Medical Museum*, conserv. L. W. G. Malcolm (London: Wellcome Historical Medical Museum, 1930).

Williams, Roger L. *Botanophilia in Eighteenth-Century France: The Spirit of the Enlightenment* (Dordrecht: Kluwer Academic, 2001).

Wittmann, Barbara. "Outlining Species: Drawing as a Research Technique in Contemporary Biology," *Science in Context, Supplement Knowledge in the Making* 26 (2013): 363–91.

Wright, John. *The Naming of the Shrew* (London: Bloomsbury Academic, 2014).

Wright, Shane Donald, and Len Norman Gillman. "Replacing Current Nomenclature with Pre-Existing Indigenous Names in Algae, Fungi and Plants," *Taxon* 71 (2022): 6–10.

Ximénez, Francisco. *Quatro Libros de la Naturaleza* (Mexico: en casa de la Viuda de Diego Lopez Daualos, 1615).

Zadeh, Lotfi A. "Fuzzy Sets," *Information and Control* 8 (1965): 338–53.

Zoller, Heinrich, and Martin Steinmann, eds. *Conradi Gesneri Historia Plantarum* (Zurich: Urs-Graf-Verlag, 1987/91).

Index

www.ingramcontent.com/pod-product-compliance
Lightning Source LLC
Chambersburg PA
CBHW071258180926
51613CB00001B/64
9781350509665